THE RE[illegible]

"There is personal excitement in all these essays because in order to write one of them I have to organize what knowledge I happen to have on the subject and flesh it out with material I can find in my reference library. In short, I educate myself, and invariably know more about any subject I deal with, after the essay is written, than I had known beforehand—and self-education is a continuing source of pleasure to me, for the more I know, the fuller my life is and the better I appreciate my own existence."—*Isaac Asimov*

Comprised of seventeen essays, THE RELATIVITY OF WRONG masterfully humanizes the enigmas of the science world, while reinforcing Isaac Asimov's unchallenged reputation as the great explainer of our age.

ISAAC ASIMOV

THE RELATIVITY OF WRONG

PINNACLE BOOKS
WINDSOR PUBLISHING CORP.

The essays in this volume are reprinted from *The Magazine of Fantasy and Science Fiction*, having appeared in the indicated issues:

"The Moon and We" (April 1986); "The Minor Objects" (May 1986); "The Second Lightest" (June 1986); "Labels on the Molecules" (July 1986); "The Consequences of Pie" (August 1986); "The Enemy Within" (September 1986); "The Relativity of Wrong" (October 1986); "The Unmentionable Planet" (November 1986); "The Dead-End Middle" (December 1986); "Opposite!" (January 1987); "Sail On! Sail On!" (February 1987); "The Incredible Shrinking Planet" (March 1987); "The Light-Bringer" (April 1987); "Beginning with Bone" (May 1987); "New Stars" (June 1987); "Brightening Stars" (July 1987); "Super-Exploding Stars" (August 1987);

PINNACLE BOOKS

are published by

Windsor Publishing Corp.
475 Park Avenue South
New York, NY 10016

Second Pinnacle Books printing: June, 1990

Printed in the United States of America

TO MY BROTHER, STAN ASIMOV,
WITH WHOM I HAVE NEVER EXCHANGED A
CROSS WORD

Contents

My Favorite Writing

I've been writing these essays at the rate of one a month for thirty years. I loved doing it at the start and my love has not diminished with the decades. I still find myself scarcely able to wait for a month to pass so that I can write another.

For one thing, I am given full liberty of choice by *The Magazine of Fantasy and Science Fiction,* which has published my essays in every issue without fail since November 1958, and by Doubleday, which has been putting out collections of the essays, in hardcover, since 1962. I am allowed to write on any subject and present it in any way I choose. Although these are science essays, I can even occasionally write an essay on a nonscientific subject if I wish, and no one objects.

Furthermore, there is no danger of my ever running out of subjects. Science is as broad as the Universe and it refines itself year by year as knowledge advances. If I write an article on superconductivity now, it is bound to be a different article from one I would have written a couple of years ago.

In fact, I include in this volume an article on the planet Pluto which I wrote a little over half a year ago. I also include a sizable addendum which includes additional information over and above what was known at the original time of writing.

There is personal excitement in all these essays because in order to write one of them I have to organize what knowledge I happen to have on the subject and flesh it out with material I can find in my reference library. In short, I educate myself, and invariably know more about any subject I deal with, after the essay is written, than I had known beforehand—and self-education is a continuing source of pleasure to me, for the more I know, the fuller my life is and the better I appreciate my own existence.

Even when my self-education falls short, and I end up getting something wrong, either through carelessness or through ignorance, my readership is such that I invariably get letters telling me of my error—always polite and sometimes even hesitant, as though the reader can't really believe I'm wrong. That sort of education is welcome, too. I may blush, but learning is worth it.

And more than that is the feeling I get that those who read my essays sometimes come to understand something they did not know before. I get a considerable number of letters telling me so. This is wonderful, too, for if I wrote only to make money, the entire effort would be merely a transaction that enabled me to pay my rent and buy food and clothing for the family. If, in addition, I am useful to my readers, if I help them enlarge *their* lives, then I have reason to live for something better than merely fulfilling the instinct of self-preservation.

Besides, let's compare science with certain other human interests—professional sporting events, for instance.

Sports stir the blood, excite the mind, create enthusiasm. In some respects, they channel competition between different portions of humanity into relatively harmless courses. Yes, there are riots after soccer games, for instance, that lead to bloodshed, but all the riots together don't equal the carnage of a minor battle, and in the United States, at least, baseball, football, and basketball go their way with nothing worse than an occasional fistfight in the stands.

I wouldn't want sports to disappear (especially baseball, which is my own special mania), for the disappearance would leave life grayer and deprive us of a great deal that may be insubstantial but that seems vital.

And yet we could, in a pinch, live without sports.

Now compare this with science. Science, rightly used, can solve our problems and do us good, as no other instrument of humanity can. It was the coming of the machine that made human slavery totally uneconomic and abolished it when all the moral preaching of well-meaning individuals accomplished little. It is the coming of the robot that may lift from the mind of humanity all the dull, repetitious jobs that stultify and destroy the human mentality. It was the coming of the jet plane and radio and television and record player that could bring to even the most ordinary of human beings the sights and sounds of human achievements in architecture and the fine arts, which in earlier ages were available only to the aristocrats and the wealthy. And so on, and so on.

On the other hand, science, wrongly used, can add to our problems and bring closer the destruction of civilization and even the extinction of the human species. I don't have to detail the dangers of the population explosion brought on to so great a degree by the advances of modern medicine, to the dangers of nuclear war, to the incredible level of chemical pollution, to the destruction of forest and lakes by acid rain. And so on, and so on.

How important science is, then, when it brings to us life and progress in one hand, and destruction and death in the other. Who is to choose how to use science? Are we to leave the choice and our own future in the hands of some elite? Or shall we participate in it? Surely if democracy has any meaning, if the American dream has any significance, we should decide that our fate should rest, to a degree at least, in our own hands.

If we feel that we should choose our own President and our own congressmen so that they can make laws only as long as they please us, why should we not also keep science under our control, and how can we do that sensibly if we do not *understand* at least a little bit about science.

Now, then, consider how newspapers and other media of information deal with sports—the quantity and detail of specialized data that they feed the public and that the public laps up with insatiable voracity. And think of the abysmal lack of significant science reporting in any but the most important and advanced newspapers. Think of the numerous columns on astrology and the dearth of information on astronomy. Think of the detailed and enthusiastic stories on UFOs and spoon-bending, and the passing reference to the latest findings concerning the ozonosphere—the former being outright charlatanry and the latter being a matter of life and death.

Under the circumstances, anything anyone can do to redress the imbalance even slightly is important. Heaven knows that for all the high quality of my readership, its absolute number is relatively low; that my own efforts to educate reach perhaps one person out of 2,500.

However, I continue to try and I continue, indefatigably, to reach out. There's no way I can singlehandedly save the world or, perhaps, even make a perceptible difference—but how ashamed I would be to let a day pass without making one more effort. I have to make my life worthwhile—to myself, if to no one else—and writing these essays is one of the chief ways in which I accomplish the task.

Part I
Isotopes and Elements

1

The Second Lightest

The first Nobel laureate I ever met and spoke to was the American chemist Harold Clayton Urey (1893–1981). It was not a happy occasion.

I had majored in chemistry as an undergraduate at Columbia University, and I obtained my Bachelor's degree in June 1939. It was my intention to continue on in graduate work, and I took it for granted that my application to do so would be accepted.

In July, however, I was turned down on the ground that I had not taken physical chemistry, which was a prerequisite for graduate work in the field. (Unfortunately, I had been pushed toward medicine by my overzealous father, and physical chemistry was not a requirement for medical school—so I had spent my time on other courses.)

I was in no mood to give up, however. When registration time came in September, I went to Columbia and insisted on an interview with the registration committee. At the head of the committee was Urey, who was head of the chemistry department.

He was one other thing, too. He was unalterably anti-Asimov. The trouble was that I was loud, gauche, irreverent, and sharptongued, and was consequently viewed with suspicion by most of the faculty. (No one doubted my intelligence but that, somehow, didn't seem to count for much.)

I asked the committee to allow me to take physical chemistry so that when I had completed the course, I could again apply for status as a graduate student. It meant losing a year, but there seemed nothing else I could do. Urey, however, required no time for consideration. As soon as I was finished asking, he said, "No!" and pointed to the door.

I had no intention of giving up, so I obtained a course catalog and found a passage that said it was possible to be an "unclassified graduate student" in order to make up a missing course, provided one fulfilled certain requirements (all of which I fulfilled). I returned the next day, waving the catalog, and repeated my request. Urey shook his head and pointed to the door again. I held my ground and demanded to know the reason for his refusal. "On what grounds?" I said.

Actually, since he didn't have grounds, except for a general dislike of me which he didn't want to admit to, he told me to come back in the afternoon. I did, and he then made me a proposition.

I would be allowed to take physical chemistry, provided I also took a full list of other courses, all of which had physical chemistry as a prerequisite. In other words, in all these other courses, the professors would assume the students already knew physical chemistry, and all of them would—except me.

Furthermore, I would be on probation, and if I did not obtain a B average, I would be dropped without credit, so that if I went to another school, Columbia would not let me have any document that would indicate I had passed certain courses, and I would be forced to repeat them. That would mean the loss of a sizable hunk of tuition money, and in those days I had no money to lose.

It is clear to me *now* that Urey was making me an offer he was sure I would not accept, so that he could be rid of me once and for all. However, he underestimated my faith in my own abilities. I accepted the offer without hesitation. After that, I eventually got my B average, was taken off probation, and went on to complete my graduate work successfully.

It has always been difficult for me, ever since, to think of Urey kindly, even though he was on my side in politics. (In 1940, when most of the faculty sported Willkie buttons, Urey's read "Roosevelt—Labor's Choice.") However, he was a top-notch scientist, whether he liked me or not, so let's go into the matter of his Nobel prize.

The story begins in 1913, when the English chemist Frederick Soddy (1877–1956) first presented strong arguments to the effect that the various atoms of a particular element need not all be identical, but could exist in two or more varieties, which he called "isotopes."

It was clear from the start that isotopes of a particular element did not differ in chemical properties. Soddy's work, however, clearly showed that they differed in mass.

Two years before Soddy's announcement, the New Zealand-born physicist Ernest Rutherford (1871–1937), with whom Soddy had worked, had put forth the notion of a nuclear atom, something which was quickly adopted by physicists. According to this notion, the atom contained a tiny massive nucleus surrounded by a number of electrons.

It was the number and arrangement of electrons that governed chemical properties, so that it was clear that isotopes of a particular element must have identical electron numbers and arrangements or their chemical properties would not be identical. That meant that the difference that distinguished isotopes had to reside in the nucleus.

In 1914, Rutherford presented his reasons for supposing that the simplest nucleus, that of hydrogen, consisted of a single particle, which he called a "proton," and that more complicated nuclei were made up of conglomerations of protons. The individual proton is 1,836 times as massive as the electron, but has an electrical charge of precisely the same size, though of opposite nature. The charge of the proton is +1, that of the electron, -1.

In an ordinary atom, which is electrically neutral, the nucleus must contain just the number of protons required to equal the number of electrons outside the nucleus. Thus, the uranium atom, which has 92 electrons outside the nucleus, must have 92 protons inside.

However, the uranium nucleus has a mass 235 times that of a proton. To get around this anomaly, the physicists of the time (for whom protons and electrons were the only known subatomic particles) assumed that in addition to protons, the nucleus would contain proton/electron pairs. A proton/electron pair would have about the mass of a proton (since the electron was so light its mass scarcely mattered). What's more, since the electric charges of the protons and electrons cancel each other, a proton/electron pair has a zero electric charge.

Therefore, it might be that a uranium nucleus would be made up of 92 protons plus 146 proton/electron pairs. The total mass would be 238 times that of a single proton, so that the "atomic weight" of uranium is 238. Since the uranium nucleus has a positive electric charge equal to that of 92 protons, the "atomic number" of uranium is 92.

Actually, it turned out that the concept of the proton/electron pair inside the nucleus didn't hold up. The pair consisted of two separate particles, and certain nuclear properties depended on the total number of particles in the nucleus. Those nuclear properties wouldn't work out properly unless the proton/electron pairs were replaced by single particles. The single particle would have to duplicate the properties of the proton/electron pair, so that it would have to have about the mass of a proton and be electrically uncharged.

Such a particle, widely hypothesized in the 1920s, was difficult to detect because of its lack of charge. It was finally discovered only in 1932 by the English physicist James Chadwick (1891–1974). He called it the "neutron," and it took the place, almost at once, of the proton/electron pair. Thus, the nucleus of the uranium atom can be viewed as being made up of 92 protons and 146 neutrons.

During the 1920s, physicists used proton/electron

pairs to explain the nature of isotopes, but in order to avoid misleading the Gentle Reader, I will speak of neutrons only, even though it is anachronistic to do so for events prior to 1932.

The nuclei of all uranium atoms *must* have 92 protons. Any deviation from that number would mean that the number of electrons outside the nucleus would have to be something other than 92. This would change the chemical properties of the atom and it would no longer be uranium. However, what if the number of neutrons changed? That wouldn't alter the charge of the nucleus or the number of electrons outside the nucleus, so that uranium would remain uranium. The *mass* of the nucleus would, however, change.

Thus, in 1935, the Canadian-American physicist Arthur Jeffrey Dempster (1886–1950) discovered uranium atoms that, in addition to the 92 protons in the nucleus, contained 143 neutrons (*not* 146). The atomic number is still 92 but the mass number is 92 + 143 = 235. Therefore, we have uranium-238 and uranium-235, and these are the two isotopes of uranium that occur in nature. They don't occur in equal quantities, to be sure, but nothing in the isotope theory suggests they must. In fact, for every atom of uranium-235 in nature, there are 140 atoms of uranium-238.

Soddy worked out his isotope concept from a detailed study of radioactive atoms and their manner of breaking down. That, however, was a weak point in his theory. Radioactivity had been discovered in 1896 and seemed to involve only very massive atoms at first, atoms that broke down spontaneously into somewhat lighter atoms. The radioactive atoms seemed very different from ordinary atoms and it might be argued that perhaps isotopes existed *only* in those radioactive elements.

Uranium (atomic number 92) and thorium (atomic number 90) were the two radioactive elements that occurred in nature to an appreciable extent, and their breakdown ended, finally, with the formation of the stable element lead (atomic number 82). However, uranium broke down to a variety of lead whose nucleus consisted of 82 protons and 124 neutrons (lead-206), while thorium broke down to a lead nucleus of 82 protons and 126 neutrons (lead-208).

If this were so, then lead must consist of these two isotopes at least, and it must exist in nature as a mixture of them in varying proportions. Lead that is extracted from thorium ores must be rich in lead-208, and have a higher atomic weight than lead extracted from uranium ores. In 1914, Soddy carefully determined the atomic weight of lead from different sources and showed that there was indeed an easily detected difference in atomic weight.

The fact that the stable element lead consisted of isotopes was not, in itself, an important broadening of the concept, because the lead isotopes result from the breakdown of radioactive elements. What was needed was some demonstration that isotopes occurred in elements that had nothing to do with radioactivity at all.

Stable elements (other than lead) do not show significant differences in atomic weight when obtained from different sources or purified by different methods. That may be either because all their atoms are alike, or because they always consist of the same mixture of isotopes.

What if one could separate the isotopes, however (assuming they are there to be separated)? An ordinary way of separating two different substances is to take advantage of differences in chemical properties. The isotopes of a particular element, however, are essentially identical in chemical properties.

However, two isotopes of a particular element are different in mass. Suppose a mixture of the nuclei of such isotopes is made to speed through electromagnetic fields. (Physicists knew how to set up such a situation in Soddy's time.) The nuclei, being electrically charged, would interact with the field and follow a curved path. The more massive nuclei have greater inertia and would therefore curve slightly less. If the nuclei in the course of their motion were made to fall on a photographic plate, the developed photograph would show a doubled curve, as each isotope followed its own slightly different path.

In 1912, the English physicist Joseph John Thomson (1856–1940) noticed such a slightly doubled path in connection with flying nuclei of the element neon. He wasn't sure as to the meaning of this, but when the isotope concept was announced the next year, there seemed a chance that what he had detected were two neon isotopes.

One of Thomson's assistants, Francis William Aston (1877–1945), set about studying the matter in earnest. He worked out a device in which the electromagnetic field caused all the nuclei of a particular mass to fall on one spot on the photographic film. The device was called a "mass spectrograph." From the position of the marks that resulted, the masses of the isotopes could be calculated, and from the intensity of the marks, the relative quantities.

In 1919, Aston was able to separate neon nuclei in such a way as to show that the element consisted of two isotopes, neon-20 and neon-22. What's more, of all the neon atoms roughly 9 in 10 were neon-20 and 1 in 10 was neon-22. That explained why the atomic weight of neon was 20.2. (In later years, as the mass spectrograph was refined, a third isotope, neon-21, was detected. We now know that out of every thousand neon atoms, 909 are neon-20, 88 are neon-22, and 3 are neon-21.)

Aston, through his mass-spectrographic work, found that a number of stable elements consisted of two or more isotopes, and this definitely established Soddy's isotope concept. Nothing has ever happened since to place it in doubt.

Whenever the atomic weight of an element is considerably removed from a whole number, we can be sure it consists of two or more isotopes, whose masses and relative quantities average out to the atomic weight.

A number of elements have atomic weights that are almost exactly whole numbers, and then it is quite possible that all the atoms of that element are indeed of the same mass. For instance, fluorine consists only of fluorine-19, sodium of sodium-23, aluminum of aluminum-27, phosphorus of phosphorus-31, cobalt of cobalt-59, arsenic of arsenic-75, iodine of iodine-127, gold of gold-197, and so on.

In the case of those elements with only one nuclear species present (there are nineteen of them), it is difficult to speak of an "isotope," since the term means there are two or more varieties of an element. For that reason, the American chemist Truman Paul Kohman (1916–) suggested, in 1947, that each atomic variety be called a "nuclide."

The term is frequently used, but I doubt if it will ever replace the word "isotope," which is already ground too deeply into the language to be removed. Then, too, physicists have learned to create isotopes in the laboratory that don't occur in nature. These artificial isotopes are all radioactive, so that they are called "radioisotopes." Any element that has only one stable nuclide is sure to have a number of radioisotopes that can be formed. There is no element that consists of only a single nuclide if one counts in the possible radioisotopes, and, therefore, strictly speaking, the term "isotope" is usable at all times. We need only say that fluorine, for instance, has only one *stable* isotope, implying the existence of radioisotopes as well.

Some elements, to be sure, have atomic weights that are very close to whole numbers and are yet made up of a number of stable isotopes. What happens in that case is that the element is made up preponderantly of one of those isotopes, with the others quite rare and, therefore, contributing little to the atomic weight.

One startling example of this was discovered in 1929. The American chemist William Francis Giauque (1895–1982) used the mass spectrograph to show that oxygen consisted of three isotopes, oxygen-16, oxygen-17, and oxygen-18, all stable. Of these, however, oxygen-16 was by far the most common. Out of every 10,000 oxygen atoms, 9,976 are oxygen-16, 20 are oxygen-18, and 4 are oxygen-17.

This shook chemists, since for a hundred years they

had been arbitrarily setting the atomic weight of oxygen equal to 16.0000 and measuring all other atomic weights against that as a standard. After 1929, this came to be known as "chemical atomic weights," while physicists used the mass of oxygen-16 = 16.0000 as the standard for "physical atomic weight." In 1961, chemists and physicists compromised by using carbon-12 = 12.0000 as the standard. That was pretty close to the chemical atomic weight table.

The oxygen = 16.0000 standard might have remained satisfactory if you could be sure that the mixture of isotopes of each element always remained precisely the same at all times and under all conditions. If the different isotopes of an element had *precisely* the same chemical properties, the mixture would always be identical, but they don't. The chemical properties are essentially the same but there are tiny differences. The more massive isotopes are always a little more sluggish about participating in any physical or chemical change than are the less massive ones. There is, therefore, the chance of finding slightly different mixes now and then.

In 1913, the American chemist Arthur Becket Lamb (1880–1952) prepared various samples of water from different sources and purified them all to an extreme. It was certain that each sample contained only water molecules with inconsiderably small quantities of any impurity. Lamb then determined the density of each sample with the utmost sensitivity of which the times were capable.

If all the water molecules were absolutely identical, all the densities should have been the same within the limits of measurement. However, the densities varied by four times the amount of those limits. It was a variation of less than a millionth from the average, but it was real, and what it meant was that all the water molecules were *not* absolutely identical. Once the concept of isotopes was introduced the next year, it could be seen that the implication was that either oxygen, or hydrogen, or both, consisted of a mixture of isotopes.

The molecule of water consists of two hydrogen atoms and an oxygen atom (H_2O). If all the water molecules contained an oxygen-18 atom, the density of such water would be nearly 12 percent higher than that of ordinary oxygen-16 water. The chances of having water that contained only oxygen-18 are virtually zero, to be sure, but small variations, depending on sources and on the methods of purification, would easily account for Lamb's results.

The fact that a massive isotope behaves more sluggishly than a less massive one opens an avenue for the separation of the two. As early as 1913, Aston had allowed neon gas to percolate through a porous partition. His feeling was that the less massive isotope (if any) would get through faster, so that the sample coming through first would be higher than normal in the less massive isotope, while the part remaining behind would be higher than normal in the more massive isotope. He repeated the procedure over and over again and eventually obtained a sample of neon which was so depleted in massive isotope that its atomic weight was 20.15 in place of the normal 20.2. He also obtained a sample of neon that had an atomic weight of 20.28 because it was enriched in the more massive isotope.

(This and other methods have been used to increase the percentage of a particular isotope in a sample of element. The most spectacular example was the enrichment procedures used to obtain uranium containing a higher than normal quantity of uranium-235, during the development of the nuclear fission bomb.)

Now there arises the question of hydrogen and its possible isotopes. Its atomic weight is just under 1.008, and that is quite close to a whole number. This means hydrogen may be composed of only a single isotope, hydrogen-1 (with a nucleus consisting of 1 proton and nothing else). If it contains a more massive isotope, that must, at the very least, be hydrogen-2 (with a nucleus made up of 1 proton plus 1 neutron) and it can only be present in trifling quantities.

Hydrogen-2 would be present in such small quantities that it was not likely to be easily detectable unless a sample of hydrogen were enriched with this more massive isotope. As early as 1919, the German physicist Otto Stern (1888–1969) tried to use Aston's diffusion method on hydrogen, but got negative results. He concluded that hydrogen was made up of hydrogen-1 only. This came about because of faults in his experimental technique, but that wasn't apparent at the time and his report discouraged further research in this direction.

Nor was the mass spectrograph of any help. To be sure there were markings that might have been the result of the presence of hydrogen-2, but these might also have been the result of hydrogen molecules made up of two hydrogen-1 atoms (H_2).

Once the oxygen isotopes were discovered in 1929, however, it became possible to determine the atomic weight of hydrogen more accurately. It seemed that the atomic weight of hydrogen was now just a little too high for it to consist of hydrogen-1 only. In 1931, two American physicists, Raymond Thayer Birge and Donald Howard Menzel (1901–76), suggested that if there were 1 atom of hydrogen-2 for every 4,500 atoms of hydrogen-1, that would be enough to account for the slightly high atomic weight.

That, apparently, inspired my future near-nemesis, Urey, to enter the field. He first attempted to detect traces of hydrogen-2 in hydrogen.

It seemed to him, from theoretical considerations, that hydrogen-2 and hydrogen-1 would give off radiation at slightly different wavelengths when heated.

Such spectral differences would be true of all isotopes, but generally, such differences would be so small as to be very difficult to spot. However, differences among isotopes increase not with the difference in mass but with the ratio. Thus, uranium-238 is three units more massive than uranium-235, but the former is only 1.28 percent more massive than the latter.

The ratio per unit difference increases rapidly, however, as the total mass decreases. Thus, oxygen-18 is 12.5 percent more massive than oxygen-16, even though the difference is only two units. As for hydrogen-2, it is 100 percent more massive than hydrogen-1, even though the difference is only one unit.

The spectral difference between the two hydrogen isotopes should therefore be much greater than that between two isotopes of any other element, and Urey felt that the ease with which the spectral difference could be detected between the two hydrogen isotopes was greater than the distinction of mass that the mass spectrograph would pick up.

He calculated the wavelength of the spectral lines to be expected of hydrogen-2 and then studied the light of heated hydrogen with a very large spectral grating. He found faint lines exactly where he thought they ought to exist.

Urey might have rushed to report this in order to gain the credit for having detected hydrogen-2, but he was a methodical and honorable scientist and realized that the very faint lines he detected might be the result of impurities in the hydrogen or of miscellaneous faults in his equipment.

The lines were faint because so little hydrogen-2 was present in the hydrogen. What he had to do, then, was to apply measures that would increase the percentage of hydrogen-2 and see if the supposed H-2 lines in the spectrum would grow stronger.

He didn't try diffusion, the method that had failed Stern. Instead, it occurred to him that if he were to liquefy hydrogen and allow it to evaporate slowly, the hydrogen-1 atoms, being less massive, would more easily evaporate than would the hydrogen-2 atoms. If he therefore began with a liter of liquid hydrogen and let 99 percent of it vaporize, the final milliliter remaining might be considerably richer in hydrogen-2 than the original hydrogen was.

This he did, and it worked. When he evaporated the final bit of hydrogen, heated it, and studied the spectrum, he found that the supposed hydrogen-2 lines had strengthened over sixfold. According to his initial calculations from all this, Urey decided that there was 1 hydrogen-2 atom for every 4,500 hydrogen-1 atoms, just as Birge and Menzel had predicted. Later work, however, showed this to be an overestimate. Actually, there is 1 hydrogen-2 atom for every 6,500 hydrogen-1 atoms.

Urey presented his results in a ten-minute talk at a meeting of the American Physical Society at the end of December 1931. His formal written reports were published in 1932.

The discovery of hydrogen-2 proved enormously important. Because of the great percentage difference between the masses of hydrogen-1 and hydrogen-2, it proved far easier to separate these two isotopes than any other two. Soon, quite pure samples of hydrogen-2 ("heavy hydrogen") were obtained, as well as samples of water with molecules containing hydrogen-2 in place of hydrogen-1 ("heavy water").

Dealing with heavy hydrogen and heavy water made the isotope seem worth a special name. Urey suggested "deuterium" from a Greek word for "second," since if all the isotopes are listed in order of increasing mass, hydrogen-1, the lightest possible, would be first, and hydrogen-2, the second lightest possible, would be second.

It was quite clear by 1934 that the eagerness with which chemists and physicists began to work with hydrogen-2 would lead to remarkable advances in science. (It did, too, as I shall explain in the next chapter) and it was not at all surprising when, in 1934, Urey received the Nobel prize in chemistry.

What's more, Urey did not rest on his laurels but went on to do important work on the origins of life, on planetary chemistry, and so on. He may not have liked me and I may not have liked him, but he was a great scientist.

2

Labels on the Molecules

A couple of weeks ago, I received a phone call from a young woman who inquired as to how she might obtain a copy of *In Memory Yet Green* (the first volume of my autobiography).

I suggested the library and she said that, indeed, she had a library copy, but the librarian was annoyed with her for continuing to renew it, and she was strongly tempted to steal it, except that this would go against her code of ethics—so what could I suggest?

There seemed no point in suggesting that she haunt the second-hand bookstores because no one but an idiot ever abandons one of my books after it has come into his possession, and there are few idiots who know enough to buy one of my books in the first place.

So I said, "Why do you want it permanently when you must have read the library copy?"

She explained, more or less as follows:

"I'm a psychologist," she said, "and I frequently

have occasion to interview adolescent boys who are having difficulties with life. I want to do a biography of your early life for junior high school students, therefore, so that I can recommend the book to the youngsters who come to me."

"Goodness," I said. "You have just told me that these adolescent boys are having difficulties. Why should you want to make things worse for them by having them read about me?"

"It wouldn't make it worse for them," she said, "it would make things better. You see, these youngsters come to me with acne. They are not athletic. They are not aggressive. They are constantly being pushed around. They bite their nails. They're afraid of girls. They can't dance. They can't ride bicycles. They're nervous in company. All they really feel comfortable with is books and homework."

"Ah," said I. "They're good in school at least, aren't they?"

"Yes, indeed," she said, "but that works against them, for they're despised for it."

"They lack a good, wholesome, all-American stupidity, do they?"

"If you want to put it that way, yes."

"Well, then, what do you do for them?"

"I tell them about you."

"About me?"

"Yes, I explain that you were exactly like them when you were young,* and now look at you—rich, famous, and successful. So if I can write a biography of you

*Not really, Gentle Reader. I *was* aggressive, and by no means nervous in company. Most of all, I was *never* afraid of girls.

angled toward the adolescent male reader, it can help them a lot, renew their hope, give them something to shoot for. You see, Dr. Asimov, *you're a role model for nerds.*"

I was speechless for a perceptible period after that, but what could I do? Here were all these adolescent boys who were seriously handicapped by their lack of stupidity and their perverted hunger for learning. Was I to leave them in the lurch?

"Come over to my place," I said, "and I will give you a copy of the book."

So she did, and I did. I signed it, too.

But after these youngsters read my autobiography, they may very well make a beeline for my writings in order to satisfy their wild craving for knowledge at the very nerdish fount.

In that case, I had better continue to churn out those writings—so here's another essay.

I'm going to carry on from where I left off in Chapter 1, where I was writing about hydrogen-2 (also known as "deuterium" or "heavy hydrogen"). I'll do so, as is often the case, by way of a digression—

We know that things change in passing through our bodies, and those changes are referred to as "metabolism," which comes from Greek words meaning (more or less) "change in passing." The air we inhale is poor in carbon dioxide and rich in oxygen; but the air we exhale is considerably richer in the former and poorer in the latter. We ingest food and drink and we eject feces and urine, while some of the food after absorption turns into bone, muscle, and other tissues while we are growing; and often into fat if we have stopped growing.

All we see with the unaided eye, however, is the starting material and the ending material, and that doesn't really tell us much if we can't see what happens in between. Seeing only the beginning and the ending gives rise to reflections like the following by the Danish writer Isak Dinesen (who was a female, despite the revered first name she adopted for her pseudonym):

> What is man, when you come to think upon him, but a minutely set, ingenious machine for turning, with infinite artfulness, the red wine of Shiraz into urine?

(This is from *Seven Gothic Tales,* published in 1934, and, if you're curious, Shiraz is an Iranian town, presumably famous for its wine in the days of the great medieval Persian poets.)

Of course, as organic chemistry developed through the 1800s, it became possible to analyze food and wastes; to realize that there were nitrogen-containing "amino acids" with molecules of a certain structure in food, and nitrogen-containing "urea" in urine, and nitrogen-containing "indole" and "skatole" in feces. All this tells us something about "nitrogen metabolism," but again mostly about the beginning and the ending. We still didn't know the vast territory in between.

In 1905 came a breakthrough, though, thanks to the work of an English biochemist, Arthur Harden (1865–1940). Along with his student William John Young, he was studying the manner in which the enzymes in yeast broke down the simple sugar glucose.

Glucose is converted into carbon dioxide and water, but the enzyme that brings this about doesn't work unless a bit of inorganic phosphate (an atomic grouping containing a phosphorus atom and three oxygen atoms) is present. Harden reasoned that the phosphorus atom was somehow involved in the breakdown and by carefully analyzing the mixture in which the glucose was breaking down, he obtained a tiny quantity of something he could identify as a sugar molecule with two phosphate groups attached to it.

This molecule is sometimes called the "Harden-Young ester," after the discoverers, but is more properly named "fructose diphosphate," and, obviously, in the breakdown of glucose, fructose diphosphate is an intermediate compound. It was the first "metabolic intermediate" to be isolated, and Harden, in this way, founded the study of "intermediary metabolism." For this and other work, Harden was awarded a share of the 1929 Nobel prize in chemistry.

Following in Harden's footsteps, other biochemists located other metabolic intermediates and, over the course of the next generation, managed to work out the course of metabolism of various important tissue constituents.

Such work was valuable, but it wasn't enough. The intermediates represented stationary mileposts, so to speak, along the road of metabolism. They were always present in small quantities since they were changed over to the next step almost as quickly as they were formed, and there was always the chance that some intermediates existed in concentrations too small to detect. Furthermore, there seemed no way of determining the details of the change from one intermediate to the next.

It was rather like watching sizable flocks of birds from so great a distance that the individual birds could not be seen. You could tell how the flock as a whole moved and changed position, but you couldn't tell what shifting and turmoil might go on within the flock.

It would help if some birds had a color pattern different from most, so that you could watch for those splotches of errant color. Or you might capture some wild birds, attach to one leg of each some little device that would send out radio signals, then release them. By observing the positions from which radio signals are received, you could study the inner workings of the flock.

In the study of metabolism, we are dealing with flocks of, let us say, glucose molecules. Vast flocks. Even a tenth of a milligram of glucose, a speck barely visible to the eye, consists of nearly a billion trillion molecules. All of those molecules, according to the chemical beliefs of the 1800s, would be exactly alike. There seemed no natural distinctions among them and chemists were at a loss for methods of introducing artificial distinctions.

The German chemist Franz Knoop (1875–1946) thought of a way, though. In 1904, he was working with fatty acids, a number of kinds of which could be obtained from the fat stored in various tissues. Each fatty acid consisted of a long straight chain of carbon atoms, and at one end of the chain was an acidic "carboxyl group'' consisting of a carbon atom, a hydrogen atom, and two oxygen atoms (COOH).

A peculiar thing about the fatty acids found in organisms is that the total number of carbon atoms in them (counting the carbon atoms in the carboxyl group) is invariably an even number. The number of carbon atoms in the commonest fatty acids is 16 or 18, but other even numbers can exist, both higher and lower.

It occurred to Knoop to attach a ''benzene ring'' to the fatty-acid chain, placing it at the end opposite to the carboxyl group. The benzene ring consists of six carbon atoms in a circle, with one hydrogen atom attached to each. It is a very stable atom grouping and is not likely to be disturbed in the body. Knoop's idea was that the benzene-attached fatty acid would meet pretty much the same fate the original fatty acid would have met, and that the final product might still have the benzene ring attached so that it could be identified. In other words, the fatty acid would have a ''label'' which would persist and would identify the end product.

It was the very first use of a labeled compound intended to elucidate a biochemical problem.

Knoop discovered that if he added labeled fatty acids to the diet, he would eventually recover the benzene ring from the fat of the animal and that a 2-carbon chain would be attached to the ring, the outer carbon being part of a carboxyl group. The name of the compound is ''phenylacetic acid,'' and Knoop obtained it no matter how long the labeled fatty-acid carbon chain he had used.

Knoop then went on to the next step, which was to make use of a fatty acid with an *odd* number of carbon atoms in the chain. These were not found in living organisms but they could be synthesized in the laboratory. They had properties that were just like those of fatty acids with even numbers of carbon atoms, and there was no obvious reason why they shouldn't occur in living tissue.

Knoop labeled the odd fatty acids with the benzene ring and fed them to animals. They didn't seem to be in any way harmed by the odd-number carbon chain, and when Knoop studied the fat, he found that the benzene ring had ended up attached to an atom group containing just *one* carbon atom, and that carbon atom was part of a carboxyl group. The compound is called "benzoic acid," and Knoop found that benzoic acid appeared no matter how long the odd-number carbon chain had been to begin with.

Here is how Knoop interpreted his findings. He decided that each fatty acid was broken down by the removal of a 2-carbon group at the carboxyl end. The cut end was then "healed" by conversion to a carboxyl group. Then another 2-carbon group was clipped off, and so on. In this way, an 18-carbon fatty acid could be clipped to 16, then 14, and so on all the way to a 2-carbon group. The final 2-carbon group could not be dealt with because it was attached directly to the benzene ring, and the body lacked the ability to clip it off the ring.

It is fair to suppose that if a fatty acid is cut down two carbons at a time, it is built up by reversing that procedure. By beginning with a 2-carbon fatty acid (acetic acid), which is known to exist in the body, and by adding on two carbons at a time, you would go from two to four, then to six, to eight, and so on. That would explain why only even-carbon fatty-acid chains were formed in tissues.

(Of course, Knoop was still working with beginnings and ends. He had not definitely located anything between. That was left for Harden the following year.)

It was an excellent and successful experiment that made sense, but there were two catches. The benzene label wouldn't work on any other important compounds, and no other labels of this sort were found. Secondly, the benzene group was unnatural, and might have distorted the normal metabolic processes, giving results that were not really accurate. Something better was needed; something that would act as a label anywhere but would be completely natural and would in no conceivable way interfere with normal metabolism.

Then, in 1913, came the discovery of isotopes, as I mentioned in Chapter 1. That meant that molecules differed among themselves in isotopic content. Consider the glucose molecule, which is made up of 6 carbon atoms, 12 hydrogen atoms, and 6 oxygen atoms. The carbon atoms can each be either carbon-12 or carbon-13; the hydrogen atoms either hydrogen-1 or hydrogen-2; and the oxygen atoms either oxygen-16, oxygen-17, or oxygen-18.

This means that there are no fewer than 25 trillion possible isotopic species of the glucose molecule, and all of them could, in theory, exist in a large enough sample of glucose.

The various species don't exist in equal numbers, however, because the isotopes themselves don't. In the case of hydrogen, 99.985 percent of the atoms are hydrogen-1; in carbon, 98.89 percent are carbon-12; in oxygen, 99.759 are oxygen-16.

This means that in the case of glucose, 92 percent of all its molecules are made up of the predominant isotopes only: carbon-12, hydrogen-1, and oxygen-16. Only in the remaining 8 percent does one find any of the comparatively rare more massive isotopes.

The rarest isotopic variety of glucose would be made up exclusively of carbon-13, hydrogen-2, and oxygen-17 (the last being the least common isotope of oxygen). This type of glucose molecule would occur, in nature, only once in every 10^{78} molecules. This means that if the entire Universe consisted of nothing but glucose, the chances would be only one in a thousand or so that even a single one of this exceedingly rare variety could be found.

Despite this vast variety of isotopic species, the situation was not improved. The isotopic species of glucose are mixed thoroughly and always in the same proportions. To be sure, different samples of glucose would, on the basis of random variations, have the various isotopic species appear in concentrations that were a bit more or a bit less than the amount a true average would call for. These variations are so small, however, in comparison with the vast number of molecules present, that they can be ignored.

But suppose you take advantage of the sluggishness of the more massive isotopes and allow carbon dioxide, for example, to diffuse through some permeable partition. Those molecules containing the more massive atoms of carbon-13 or oxygen-18 would lag behind. If you repeated the diffusion over and over, you would end up with samples of carbon dioxide that were rich in carbon-13 (and, to a lesser extent, oxygen-18). In the same way, the boiling or electrolysis of water would leave you with samples high in hydrogen-2, while the treatment of ammonia can give you samples high in the rare isotope nitrogen-15. (The common nitrogen isotope is nitrogen-14.)

Of these four elements, by far the most important to the chemistry of life, nitrogen-15, is 7.1 percent more massive than nitrogen-14; carbon-13 is 8.3 percent more massive than carbon-12; and oxygen-18 is 12.5 percent more massive than oxygen-16. Compare this with hydrogen-2, which is 100 percent more massive than hydrogen-1.

It followed then, after the discovery of hydrogen-2, that that isotope soon became available for metabolic experimentation. It was the first isotope to be so available, but afterward, as separation techniques were refined, other comparatively rare isotopes became available, too.

In 1933, a German biochemist, Rudolf Schoenheimer (1898–1941), emigrated to the United States. (He was Jewish, and saw no percentage in remaining in Germany after Adolf Hitler had come to power that year.) In the United States, he obtained a post at Columbia University and had a chance to work closely with Urey and, therefore, to obtain a supply of hydrogen-2.

It occurred to Schoenheimer that hydrogen-2 could be used to label organic compounds. Unsaturated fatty acids, made up of molecules containing less than the maximum amount of hydrogen atoms, have the capacity to add two hydrogen atoms (or four, or six, depending on the degree of unsaturation) and become saturated. It would not matter to the unsaturated fatty acids whether they took on hydrogen-1 or hydrogen-2 and the end product could be rich in hydrogen-2.

Hydrogen-2 occurs in nature, and a particular saturated fatty-acid molecule might, therefore, contain one or more of this isotope in the ordinary course of things. Since there is roughly 1 hydrogen-2 atom present for every 6,500 hydrogen-1 atoms—and there are 36 hydrogen atoms in the typical fatty-acid molecule—there would be 1 fatty-acid molecule out of every 180 that contained 1 hydrogen-2 atom, 1 out of every 32,000 that contained 2, and 1 out of every 5,750,000 that contained 3.

This isn't much and it is easy to flood a rat's food with "isotopically labeled" fatty acids containing more hydrogen-2 than is contained in its entire fat supply. You then follow the label. After the rat has digested, absorbed, and metabolized the fat, it can be killed and its fat can be separated into its different fatty acids. These can be oxidized to carbon dioxide and water, and the water can be analyzed by mass spectrograph to determine its hydrogen-2 content. Anything above a very small natural amount had to be derived from the labeled fat fed the rat.

Beginning in 1935, Schoenheimer, in collaboration with David Rittenberg (1906–   ), began a series of such tests on rats.

An animal, eating food, would absorb portions of that food into its body, would use some to build up its own tissues, and would oxidize other portions in order to obtain energy with which to carry on its various functions. Any portion of the food left over could be stored away as fat, serving as an energy reservoir until those times when the animal could not find enough to eat.

Why fat? Because fat represented the most compact form in which the animal body can store energy. A given quantity of fat, on oxidation, releases more than twice the energy that the same quantity of carbohydrates or proteins would.

It was assumed, as a matter of course, that these reserve stores of fat were relatively immobile; that the fat molecules were just waiting for emergency use. Since the animal might be fortunate enough to encounter food shortages only rarely, or even never, the fat stores might rarely or never be called on, and the molecules would then just lie there and slumber peacefully, so to speak.

But they don't. After Schoenheimer and Rittenberg had fed the rats the isotopically labeled fats, they waited four days, then analyzed the stored fat of the rat body. They found that half the hydrogen-2 atoms the rats had eaten were in that fat store. What this meant was that the rat (and, presumably, any other animal, too) was constantly using molecules from the fat store and replacing them with other molecules, or else that the molecules of the fat store were constantly exchanging hydrogen atoms with each other and with new molecules that arose. In either case there was rapid and incessant activity.

Schoenheimer and Rittenberg tried other types of isotopic labels as well. They obtained a supply of nitrogen-15 from Urey and used it to synthesize amino acids. Amino acids are the building blocks of protein molecules and there is at least one nitrogen atom in every amino acid. An amino acid labeled with nitrogen-15 can be fed to rats and that label can be followed.

It turned out that the nitrogen atom did not remain in the particular amino acid fed the rats. After a remarkably short period of time, it was found in other amino acids.

This turned out to be a general rule. The constituent molecules of the body are not just sitting there waiting for some signal that a chemical change involving them is needed. Instead, they are constantly reacting.

These reactions need not imply any overall change, of course. A molecule could give up a pair of hydrogen atoms and pick them up again. It could give up the constituent atoms of a water molecule and pick them up again. It could give up a nitrogen-containing group and pick it up again. A molecule with a ring of atoms might break the ring and reform it, while a molecule with a straight chain of atoms might form a ring and then break it. Two molecules might exchange identical atoms or atom groups, leaving each in the same state as before.

None of this could have been demonstrated without the use of isotopically labeled compounds, but once such labels came into use and demonstrated this rapid, endless molecular change, one could see clearly (in hindsight) why this should be.

If molecules remain quiescent, unmoving, inert, and if they merely waited for some emergency, then when the emergency came there would have to be some drastic change in the molecular environment to convert quiescence into action. It would undoubtedly take time to "rouse" the molecules and get all the machinery into shape. The end result would be that it would be very unlikely that the organism could meet the emergency quickly enough.

If, on the other hand, molecules were always in action, quivering in place (so to speak), then, in case of emergency, only minor changes would have to be made. The molecules, rapidly undergoing a variety of changes in any case, would merely have to accelerate some and slow others. All the machinery would, so to speak, be already in place.

If there were organisms in Earth's early history that didn't make use of ever-active molecules (which, somehow, I doubt), they would have quickly been nosed out in the evolutionary race when other organisms that *did* have ever-active molecules developed.

Schoenheimer published a book entitled *The Dynamic State of Body Constituents* in which he described and interpreted all his findings, and it made a splash in the chemical and biochemical world. But then, on September 11, 1941, when he was forty-three years old, he committed suicide.

I don't know why he did this. To be sure, he had fled Hitler and, in September 1941, it seemed that Germany was winning. All of Europe was under its control. Great Britain had barely survived the air blitz and the Soviet Union, newly invaded, seemed to be collapsing under the powerful German offensive. Japan was on the Nazi side and the United States was immobilized by the pressure of its own isolationists. I well remember the fear and depression of that time for anyone who had cause to dread the Nazi racial theories. Schoenheimer might have had personal reasons, too, but I can't help but feel that the state of the world contributed.

In any case, it was a tragedy in several ways. Consider that Schoenheimer had founded the technique of isotopic labeling and had, in the process, revolutionized our view of metabolism. Consider, too, that further work (in which Schoenheimer would undoubtedly have participated had he lived) made use of such tracers and solved many metabolic problems in detail. It would certainly seem, then, that in a very few years, Schoenheimer would surely have received a Nobel prize, had he been able to let himself go on living.

What's more, he didn't live to see the heyday of another form of isotopic label that came in after World War II was over. He himself, if it were possible for him to know that, might have bemoaned that loss even more than the loss of the Nobel prize.

We'll take up that other form of labeling in the next chapter.

3

The Consequences of Pie

On November 11, 1985, my doorman said to me as I came in, "You are on Page Six of the New York *Post,* Dr. Asimov."

My eyebrows went up. Page Six is the page on which personal items are retailed—a kind of gossip page. At least, so I'm told; I don't often see the *Post.* "What about?" I said.

The doorman grinned. "You were kissing a woman, Dr. Asimov." And he handed me the paper.

Well, my kissing a woman isn't news. Personally, I think that women have been particularly put together for kissing. So why should the *Post* bother? I opened the paper to Page Six as the elevator took me up to my apartment.

I walked into the apartment and said to my dear wife, Janet, "It's finally happened, Janet. I kissed a woman and it made the gossip columns in a newspaper."

"Oh no," said Janet, who knows all about this amiable weakness of mine. "Now everyone we know will phone to tell me about it."

"Who cares?" I said, and I handed her the paper. Here is the item, complete:

> A city kid like Isaac Asimov doesn't need a drive-in movie. The prolific sci-fi [sic] writer didn't seem to care who saw him hugging and kissing a woman at the New York Academy of Science on East 63rd Street during a recent screening of the new TBS show, *Creation of the Universe.* And why should he? The lady was his wife of 12 years, Janet Jeppson. Maybe it was the title that got the two sexagenarians going.

Janet laughed very heartily. She was so amused, she didn't even mind being called a sexagenarian, even though at the time of the incident (November 5), she was only 59 ¼ years old.

But I said, "You miss the point, Janet. Think what it tells us of our society. We have a man in his late youth kissing his own wife, and this is considered so unusual that it makes the newspapers."

And yet odd items are not only recorded in newspapers, but even in history books—and the most trivial things might turn out to be important. In the history of science, for instance, there is the sinister incident of the landlady and her Sunday pie—

The story involves a Hungarian chemist named Gyorgy Hevesy (1885–1966). His father was an industrialist ennobled by the Austro-Hungarian Emperor, Francis Joseph I, so that the chemist's name is sometimes given as "von Hevesy."

In 1911, Hevesy had a dispute with his landlady. He claimed that the remains of the pie she routinely served on Sunday were recycled and put into the food served during the remainder of the week. (For myself, I don't consider that a criminal operation, but in the days before the commonplace presence of efficient home refrigeration, such recycling might have been risky.) Naturally, the landlady denied the charge vigorously.

As it happened, Hevesy was working in Ernest Rutherford's laboratory at Cambridge at the time. Rutherford and his students were deeply engaged in research on radioactivity and this meant that Hevesy could get a small trace of radioactive substance. What he used, in fact, was a tiny bit of the breakdown products of thorium.

When the meal was finished one Sunday, Hevesy added a smidgen of the radioactive substance to the pie when no one was looking. The following Wednesday, a soufflé was served, and Hevesy got out his electroscope.

An electroscope consists of two gold leaves enclosed in a chamber. The gold leaves are attached to a rod, one end of which sticks out of the chamber. If the outer end of the rod is touched by an electrically charged object, the two gold leaves are each similarly charged and repel each other, so that they form an upside-down *V*.

If such a charged electroscope is subjected to hard radiation of the kind that radioactive substances produce, then the charge is carried away and the two leaves collapse toward each other. When the electroscope was brought near the soufflé, the gold leaves at once began to collapse. In other words, the soufflé was radioactive and it was so only because it contained bits of the Sunday pie.

Hevesy had, in other words, marked the pie with a radioactive label, and had then traced the movements of that label. It was the first use of a "radioactive tracer" in history, albeit for a trivial purpose.

Hevesy himself disparaged the event and considered it unimportant, but that can't be so. If nothing else, it got him to thinking about radioactive tracing, and that had its consequences.

In 1913, he applied the principle of radioactive tracing to a chemical problem. Many lead compounds are only slightly soluble. It is of chemical interest to know just how soluble each might be, but it is hard to make accurate measurements of the matter. Suppose you powder a lead compound and add it to water. You stir it till as much of the compound has dissolved as possible. You then filter off the undissolved powder and analyze the clear fluid for dissolved compound. There is so little compound present, however, that it is very difficult to determine its concentration exactly.

Hevesy decided that it was only necessary to mix ordinary lead with lead-210, which is formed in the course of uranium breakdown and was, in those days, called "Radium D." Lead-210 would mix with ordinary lead and, since its chemical properties were identical with ordinary lead, it would undergo whatever changes the latter did. The lead with its radioactive admixture would then be used to form a particular compound, which would contain a tiny percentage of lead-210. The exact quantity of lead-210 present could easily be determined by measuring the intensity of radioactive radiation. This is such a sensitive type of measurement that it would yield accurate results despite the small quantity present.

If the lead compound were now dissolved, the lead-210 version of the compound would dissolve also, and in precisely the same proportion as the compound itself. By measuring the percentage of the lead-210 that was present in solution, we would automatically be measuring the percentage of the total compound that had dissolved. In this way, solubility could be determined far more accurately than by earlier methods.

By 1918, Hevesy was using both radioactive lead and radioactive bismuth to study the behavior of hydrogen compounds of those metals.

Then, in 1923, Hevesy used radioactive tracers, for the first time, in biochemical research. He added small amounts of a lead solution to the fluid used to water plants he was working with. Plants take up mineral salts from water in the soil, and, presumably, they would take up lead compounds, too, in very small quantities. Hevesy had used lead compounds, for the purpose, that had a bit of radioactive lead-210 present. At various intervals, plants were burned and the ash was analyzed for radioactivity. In this way, the progress of lead uptake and the lead content of various portions of the plant could be followed accurately.

There is, however, a limit to what you can do with lead and bismuth, especially in biochemical problems, since neither element occurs naturally in living tissue (except as an accidental contaminant). For that reason, Hevesy's reports, while they seemed to have some interest, were regarded as a dead end. It was not until 1943 that the consequences of his work (and of his landlady's Sunday pie) were seen to be supremely important, and he was then given the Nobel prize in chemistry.

Here's how radioactive tracing came to assume great importance—

At first glance, it would seem that radioactivity is confined entirely to the exotic elements at the upper end of the periodic table. Uranium (element number 92) and thorium (number 90) break down and produce dozens of different products. These products include atoms with an atomic number as low as 52, but no lower. (There were far too many breakdown products for each to have a separate atomic number and it was this that first placed Frederick Soddy on the track of isotopes—as I mentioned in Chapter 1.)

Of all the breakdown products only those that were isotopes of lead (number 82) or bismuth (number 83) were of elements that also possessed stable isotopes. The study of radioactive phenomena through the 1920s revealed no radioactive isotopes of any element with an atomic number less than 82, and it seemed a reasonable supposition that radioactive isotopes of these lighter elements simply didn't exist.

Then came the work of Frederic Joliot-Curie (1900–58) and his wife, Irene Joliot-Curie (1897–1956), who was the daughter of the famous Madame Marie Curie.

The Joliot-Curies were busily engaged in bombarding such light atoms as boron, magnesium, and aluminum with alpha particles, a type of radiation produced by some radioactive substances. This sort of work had been initiated by Rutherford, who was the first to observe that atomic nuclei were altered as a result.

An alpha particle is made up of 2 protons and 2 neutrons, and when it strikes the nucleus of a light atom, it can happen that the 2 neutrons along with 1 of the protons will remain in the nucleus while the other proton flies away. Rutherford noted this first in 1919, when he bombarded nitrogen with alpha particles. The nitrogen nucleus has 7 protons and 7 neutrons. If 1 proton and 2 neutrons from the alpha particle are added, you have a product containing 8 protons and 9 neutrons.

A nucleus with 8 protons and 9 neutrons is that of oxygen-17, which is rare in nature, but stable. Thus, Rutherford had converted nitrogen-14 to oxygen-17 and had accomplished the transmutation process of changing one element into another that had eluded the early alchemists.

The Joliot-Curies obtained similar results. In 1933, they found that bombarding aluminum-27 (13 protons and 14 neutrons in the nucleus) with alpha particles (which have 2 protons and 2 neutrons) resulted in the addition of 1 proton and 2 neutrons to the nucleus, producing one with 14 protons and 16 neutrons. This is the nucleus of silicon-30, a rather rare but stable isotope of silicon.

Naturally, this meant that, as usual, protons were ejected by the bombarded aluminum. That was not surprising at all. But then the Joliot-Curies noted that, in addition to the protons, a certain quantity of neutrons and positrons were also being emitted. This was a little more surprising, but not too much so.

A neutron (which had been discovered in 1931, only four years earlier) is very similar to a proton, except that the neutron has no electric charge while the proton has a charge of +1. The positron (discovered only two years earlier) is very light compared to either the proton or the neutron and, like the proton, has a charge of +1. Combine the neutron and positron and you have a particle that is still about the mass of a neutron and has a charge of +1. In short, you have a proton. Therefore, if, as a result of a nuclear reaction, a proton is formed, it is conceivable that, in the same nuclear reaction, a neutron plus a positron, which, together, are the equivalent of a proton, may also be formed.

So far, so good. In early 1934, the Joliot-Curies noted that when the alpha particle bombardment stopped, the production of protons and neutrons stopped also, and at once. That was to be expected. However, here came the big surprise. The production of positrons did *not* stop. It kept right on going, at a rate that diminished with time in a way characteristic of a radioactive transformation.

What was happening?

The Joliot-Curies had supposed, at the start, that the aluminum atom ejected a neutron and a positron at the same time, and that since this was equivalent to ejecting a proton, the aluminum-27 was changing to silicon-30, either way. The fact that the neutrons stopped being ejected and the positrons kept on coming might well mean that the two particles were produced independently. Suppose, then, that a neutron was produced and ejected first.

This would mean that when the alpha particle struck the aluminum-27 nucleus, 2 protons and 1 neutron were absorbed from the particle, while the second neutron was ejected. The 13 protons and 14 neutrons of aluminum-27 would thus be converted into 15 protons and 15 neutrons and that would be a nucleus of phosphorus-30.

Phosphorus-30, however, does *not* exist in nature. Phosphorus atoms that do exist in nature occur in only a single atomic variety—phosphorus-31 (15 protons and 16 neutrons). No other isotope of phosphorus exists in nature.

Yet suppose phosphorus-30 were formed. It would have to be radioactive, for that would explain why it doesn't occur in nature. Even if it were somehow formed, it would break down speedily.

In fact, what if the phosphorus-30 breakdown was accompanied by the ejection of positrons? That would explain why positrons kept on being ejected after the alpha-particle bombardment had ceased. The alpha-particle bombardment formed phosphorus-30 faster than it could break down so that a certain small concentration would be built up. Then, after the bombardment stopped, the built-up phosphorus-30 would continue to break down.

From the rate at which positron formation fell off, it could be calculated that the half-life of phosphorus-30 is about 2.5 minutes.

Positron ejection is very much like beta-particle ejection. Beta particles are speeding electrons, after all, and a positron is exactly like an electron, except that the former has a charge of +1 and the latter one of −1.

When an electron is ejected from a nucleus, a neutron with a charge of 0 is converted into a proton with a charge of +1. In other words, to have a nucleus lose a negative charge (by emitting an electron) is equivalent to saying it has gained a positive charge (by converting a neutron to a proton).

Positron ejection would naturally do the reverse of electron ejection, since a positron is the opposite of an electron. If the ejection of an electron turns a neutron into a proton, the ejection of a positron turns a proton into a neutron. If phosphorus-30 ejects a positron, then its 15 protons and 15 neutrons change into 14 protons and 16 neutrons and it becomes silicon-30.

What it amounts to, then, is that if aluminum-27 is bombarded with alpha particles, it can change into silicon-30 directly, or it can change into silicon-30 indirectly by way of phosphorus-30. The Joliot-Curies, therefore, were the first to demonstrate the existence of "artificial radioactivity." The importance of this was recognized at once and, in 1935, they were awarded the Nobel prize in chemistry.

Once the Joliot-Curies had shown the way, other investigators followed in their track. Radioactive isotopes (or "radioisotopes") were found in large numbers and, eventually, every single element on the list, with no exceptions, was found to have radioisotopes.

Obviously, radioisotopes are likely to make better labels than do stable, but rare, isotopes. A stable isotope can only be detected, and its concentration measured, by mass spectrometry, which is rather tedious and difficult. Radioisotopes can be detected, and their concentration measured, much more quickly and easily.

Hevesy was the first one off the mark here, too. In

1935, he studied the uptake by plants of phosphate ions from solution, using radioactive phosphorus as the tracer.

Of course, there are difficulties involved in using radioisotopes. What if the half-life is short?

As I said, phosphorus-30 has a half-life of 2.5 minutes. Obviously, any experiment using phosphorus-30 must be completed from beginning to end in a few minutes, or the phosphorus-30 will have dwindled to the point where it is impossible to detect with sufficient accuracy. Fortunately, phosphorus-32, another radioisotope of the element, has a half-life of 14.3 days, and that is much better.

From the standpoint of biochemistry, the five most important elements are hydrogen (number 1), carbon (number 6), nitrogen (number 7), oxygen (number 8), and sulfur (number 16). For sulfur, there is the convenient radioisotope sulfur-35, with a half-life of 87 days.

Hydrogen seemed a more puzzling problem. In fact there might be reason to suppose that even if all the other elements had radioisotopes, hydrogen might not. After all, it is the simplest of the elements. How could it break down?

In fact, the common hydrogen nucleus is made up of 1 proton and nothing else. It would have to be stable. Even when hydrogen-2 (deuterium) was discovered, with a nucleus of 1 proton and 1 neutron, that was stable, too.

Once deuterium was discovered, however, it was

used by scientists in a number of different ways. For one thing it could be used for neutron bombardment.

Neutrons are electrically uncharged and cannot be accelerated as charged particles can be. That means that if you have a source of neutrons, you must take them at the energies with which they are produced, since you can't accelerate them to higher energies. Usually, the energies you find are not the ones experimenters wanted.

A deuterium nucleus, or "deuteron," made up of 1 proton and 1 neutron *can* be accelerated, since it has a charge of +1. Atomic nuclei can therefore be bombarded with speeding, energetic deuterons.

As it happens, though, the proton and neutron in the deuteron are held together weakly in comparison with the bonds present in other nuclei. As a speeding deuteron approaches a nucleus, then, the nucleus (which is positively charged) repels the proton half. The bond between the proton and the neutron may then be broken, so that the proton may be forced away from the nucleus and sent speeding off in a different direction. The neutron, however, being uncharged, would be unaffected by the electric charge of the nucleus, and would continue speeding forward. It might then strike the nucleus and merge with it.

In 1934, an Australian physicist, Marcus Laurence Elwin Oliphant (1901-   ), bombarded deuterium itself with speeding deuterons. Every once in a while, the proton of the deuteron was forced to break away while the neutron traveled onward, striking the deuterium nucleus (a low-energy deuteron) and remaining. The result is a nucleus with 1 proton and 2 neutrons.

This is "hydrogen-3," or, as it is often called, "tritium." Oliphant is its discoverer.

Hydrogen-3, it turned out, is radioactive, and it is the only known radioisotope of hydrogen. It breaks down by ejecting an electron (a beta particle), so that within its nucleus, a neutron turns into a proton. The resulting nucleus, with 2 protons and 1 neutron, is helium-3, an extremely rare nucleus, but stable.

The half-life of hydrogen-3 is 12.26 years, so it can easily be used as a radioisotopic label.

The luck that biochemists have had with sulfur and hydrogen failed them, however, with oxygen and nitrogen.

The least unstable nitrogen radioisotope is nitrogen-13 (7 protons and 6 neutrons), which has a half-life of only ten minutes. The situation in connection with oxygen is even worse. The most nearly stable oxygen radioisotope is oxygen-15 (8 protons and 7 neutrons) and its half-life is only about two minutes.

Neither one is very useful as a tracer, since they are too evanescent. What's more, it is as certain as anything can be that we will never find a radioisotope of either oxygen or nitrogen that will have a longer half-life. For those two elements, then, we are forced to stick with the rare, stable isotopes, oxygen-18 and nitrogen-15, as labels. (There's no point in complaining, though. We are lucky to have them, and they have served biochemists well.)

For a while, it didn't seem that carbon, *the* most important element in biochemistry, would be any better. During the 1930s, the least unstable carbon radioisotope known was carbon-11 (6 protons and 5 neutrons), which had a half-life of 20.4 minutes.

This was short, but for carbon's sake, biochemists did their best to work with it. They devised experiments that could be completed within the hour. There were certain advantages to this. If a short experiment is successfully designed, it can be repeated over and over again, sometimes under varying conditions, without too much loss of time. Then, too, a short-lived radioisotope produces copious radiation (that's why it's short-lived), so an exceedingly small amount is all that needs to be used. Still, even though some successful work was done with carbon-11, the opportunities were limited.

It was known that carbon-14 ought to exist and that it should be radioactive. Among the lighter elements, there is only one stable isotope for any given total number of protons and neutrons in the nucleus. Nitrogen-14 (7 protons and 7 neutrons) is stable, so carbon-14 (6 protons and 8 neutrons) was sure to be unstable. It was expected to break down by emitting an electron and changing a neutron to a proton. That would produce nitrogen-14.

The only argument was over what the half-life of carbon-14 might be. During the late 1930s, chemists felt that the half-life might be of the order of fractions of a second. They kept trying to isolate some form of radioactive breakdown that could be attributed to carbon-14 and kept failing. With every failure, it seemed more certain that carbon-14 must be very short-lived and that that was why it couldn't be isolated.

Then, in 1939, a Canadian-American biochemist, Martin David Kamen (1913–   ), set about painstakingly investigating every nuclear reaction that might possibly give rise to carbon-14. He made use of bombardments in which protons, deuterons, or neutrons were the bombarding particles, and boron, carbon, or nitrogen were the atoms being bombarded.

Until early 1940, the results were negative, and then Kamen bombarded carbon with deuterons of a particular energy and obtained a weak radioactivity. The radioactivity accompanied the carbon in all its chemical changes, and therefore had to involve a carbon isotope.

To have produced a carbon isotope, the deuteron would have had to add its neutron to the carbon nucleus and have its proton go its own way. An added neutron would leave the element changed but increase its mass number by 1. Thus, carbon-12, the common isotope, would be changed to carbon-13, which is also stable, though rare. Carbon-13 itself, however, would be changed to radioactive carbon-14.

If this were so, it would be best to increase the quantity of carbon-13 in the carbon being bombarded. This was done, and when the enriched carbon was bombarded with deuterons, the radioactivity was much strengthened. At last, carbon-14 was obtained in quantities large enough to be studied, and a virtual shock wave went through the world of biochemistry. Carbon-14 turned out to have a half-life of some 5,730 years!

With carbon-14, experiments could be carried out that would last for a lifetime, if one wished, and there would be no problem in dealing with the radioactivity. It would not vanish and it would remain almost constant, in fact.

There was still a catch, though, even as the 1940s opened. Radioisotopes could only be formed in small quantities and they were therefore very expensive. However, even as carbon-14 was discovered, scientists were working on uranium fission and by the end of World War II, nuclear reactors had been devised.

A nuclear reactor is a source of vast numbers of slow neutrons produced by fissioning uranium atoms. These slow neutrons are easily captured by atoms of many types and elements of higher mass number are thus formed. Or, a neutron may be absorbed and a proton or an alpha particle may be ejected, so that a radioisotope of another element might be formed. In this way useful radioisotopes for any of the elements of biochemical significance can be formed, including hydrogen-3 and carbon-14—and tracer work entered its golden age.

Carbon-14 was, of course, the most important of the radioisotope tracers, and an example of its triumphs came in connection with photosynthesis, but I'll leave that for some other day. Instead, I will take up, in the next chapter, two other important aspects of carbon-14 that do not involve ordinary tracer work at all.

4

The Enemy Within

Lester del Rey is a top-notch science fiction writer, editor, and critic. He is one of the most straightforward, the most honest, and the most intelligent people I know. He is also, I am glad to say, one of my oldest friends. I've known him for forty-five years.

In that interval, of course, he has grown forty-five years older and I have grown four or five years older myself.

Our relationship is a peculiar one. If we two are alone, there is nothing between us but warmth and friendship. As soon as a third person shows up on the horizon, however, things instantly change. Lester bares his teeth and lets me have it.

As I have said, over and over, "Lester will give me the shirt off his back. What he won't give me is a kind word."

Of course, you mustn't get the wrong idea. I give as good as I get. I'm waiting for him, someday, to say, "Here, Isaac. Here's the shirt off my back."

In which case, I'm going to say, and I can hardly wait, "Off *your* back, who would want it?"

Anyway, we were taping a television interview together some years ago, and we both talked sensibly and maintained the utmost decorum. You would think we were each of us entirely respectable.

And then, as the program approached an end, the nice woman who was interviewing us turned to me and said, "I understand, Dr. Asimov, that you don't fly. It seems strange that someone who travels all over the Galaxy in his imagination should not fly. Why is that?"

I am terribly tired of that question, but I answered, in a civil manner, "It's simply an irrational fear."

Whereupon Lester, who had controlled himself for nearly half an hour, broke down and said, "Otherwise known as cowardice. As for me, I'm ready to fly at any time."

At which, totally forgetting we were on television, I shot back at him, "That's because your life is worth *nothing,* Lester."

With that, the program was over, and the young woman, grinning, thanked us both. A wave of sick feeling came over me, however, as I suddenly realized, very clearly, that we were going to be on television that night and my dear wife, Janet, was going to be watching.

As it happens, she is very fond of Lester. I thought, rather nervously, that I had better break the news to her diplomatically.

So I called her up and explained what had happened.

She said, appalled, "You said that on *television!*" And

then, because she has a heart as soft as chinchilla fur, she began to weep.

I called Lester over and said, "Please, Lester, tell her you didn't mind."

Lester tried very hard, but she wouldn't stop and the whole next day she kept looking at me and saying, "You said it on *television.*"

I was finally desperate enough to make a logical point. I said, "Well, Lester started it."

And she trumped my ace with a logical remark of her own. "That's no excuse," she said.

—Well, now that I've thought of that episode (I just talked to Lester on the phone, and that reminded me), I'd better get my mind off it. I'll talk about carbon-14.

I've been talking about isotopic tracers in one way or another for three consecutive chapters so far and this will make the fourth. In Chapter 3, I told you about the unexpected discovery that carbon-14 was a rather long-lived radioisotope, with a half-life of 5,730 years.

Since the half-life is that long, and since carbon is the element most centrally involved in life, you can see that carbon-14 became, at once, the most important tracer in biochemistry.

If the half-life was all there was, however, there would be no carbon-14 in the natural environment today, even though a half-life of 5,730 years is long when compared to a human lifetime, or even when compared to the history of civilization.

Writing was invented about 3000 B.C. If a pound of carbon-14 were placed under the first bit of clay into which cuneiform was incised, and both objects were left undisturbed to the present day, then about half a pound of carbon-14 would still be in existence today.

The half-life, however, is not long compared to geologic ages. If the entire Earth were a solid mass of carbon-14, every bit of it would break down, to the last few atoms, in just about a million years; and a million years is only $1/4{,}600$ of the lifetime of the Earth. If, then, carbon-14 were formed, in any way, more than a million years ago, then, no matter how much of the isotope had then existed, there would be none left now.

We know of no way in which carbon-14 might be formed in Earth's past that would not be operative today as well. Therefore, if there is no natural way in which carbon-14 is formed on Earth now, there was no natural way in which it was formed at any time, and no carbon-14 should exist on Earth except for the tiny quantities that scientists can make in the laboratory.

But there *is* a small quantity of carbon-14 in nature; and this can only be because some process is manufacturing it *right now.*

The Latvian-American chemist Aristid V. Grosse (1905-), suggested in 1934 that cosmic rays interact with the atoms of the atmosphere to produce nuclear reactions that might result in the production of radioisotopes without human intervention.

Investigation eventually showed that this was so. The cosmic-ray particles that enter the upper atmosphere (the "primary radiation") are the positively charged nuclei of atoms, speeding along at about 99 percent of the speed of light. About nine tenths of the particles are the nuclei of hydrogen atoms, that is, simple protons.

The protons (and the scattering of more massive nuclei) sooner or later collide with atoms, and do so very energetically because of their speeds. The nuclei they collide with are smashed, and produce particles of "secondary radiation," somewhat less energetic than the primary radiation, but still energetic enough. Among the particles of this secondary radiation are neutrons.

Every once in a while, one of these neutrons strikes a nucleus of nitrogen-14 (the major component of the atmosphere). The neutron knocks a proton out of the nucleus, while remaining in the nucleus itself. The nitrogen-14 nucleus is made up of 7 protons and 7 neutrons. If a proton leaves as a neutron enters, the result is a nucleus with 6 protons and 8 neutrons, and that is carbon-14. For convenience, we might also call it "radiocarbon."

Radiocarbon, once formed, quickly combines with oxygen and the result is radiocarbon dioxide.

Naturally, the carbon-14 atoms in the radiocarbon dioxide break down eventually. Inside the carbon-14 nucleus, a neutron changes into a proton. A beta particle (a speeding electron) is emitted and the nucleus becomes nitrogen-14 again. In the process, the nitrogen atom is torn from the oxygen and we're back where we were before the cosmic rays struck.

Meanwhile, however, the cosmic-ray particles are producing more neutrons that are converting more nitrogen-14 to carbon-14. An equilibrium is reached in which just as many carbon-14 atoms are being formed as are being broken down. The total amount of carbon-14 atoms in the atmosphere (in the form of radiocarbon dioxide) then remains constant.

The equilibrium amount of carbon-14 in the atmosphere is very small, but radioactivity is easy to detect and that amount can be measured. It seems that 1 out of every 540 billion carbon atoms in the atmosphere is carbon-14.

That certainly doesn't sound like much, but the Earth has a lot of atmosphere. Even though very little of it is carbon dioxide, and only part of the carbon dioxide is carbon, and though only a very occasional carbon atom is carbon-14, there are still about 1,300 kilograms (or nearly 1½ tons) of carbon-14 in the atmosphere.

Nor is all of the Earth's content of carbon-14 in the atmosphere. Some carbon dioxide is dissolved in the ocean, and with it some radiocarbon dioxide is also dissolved.

What's more, plants absorb carbon dioxide as the raw material out of which their tissues are built up. Naturally, they absorb radiocarbon dioxide along with ordinary carbon dioxide, since carbon-14 has chemical properties that are identical to those of stable carbon-12 and carbon-13.

Then, too, animals eat plants and incorporate plant constituents into their own tissues, including any carbon-14 that is there. In the end, there is carbon-14 in all life forms, without exception.

Slowly, the carbon-14 in living tissue breaks down but, slowly, new carbon-14 atoms enter from the atmosphere (in the case of plants) or from food (in the case of animals). Therefore, the carbon-14 in living tissue remains constant in concentration—at least while the tissues are alive.

Once an organism dies, however, it can no longer take in carbon-14 from either the atmosphere or food.

It is stuck with whatever carbon-14 it had in its tissues at the time of death, and that carbon-14 slowly and inexorably breaks down.

We know precisely the rate at which carbon-14 breaks down, and we can detect and recognize the beta particles it gives off. From the number of beta particles we can count, we know the amount of carbon-14 in a given sample of a dead remnant of something that was once living. By comparing the amount with that contained in live matter, we can calculate how long the carbon-14 has been breaking down, and, therefore, how long the dead material has been dead.

Naturally, this doesn't work where dead organisms are eaten, and their carbon-14 is absorbed into the tissues of the eater (where the eater may be anything from a blue whale to a decay bacterium). There are, however, some dead remnants that remain intact for thousands of years. There is old wood, charcoal from old campfires, old textiles, remains of old seashells, and so on.

In 1946, the American chemist Willard Frank Libby (1908–80) suggested that carbon-14 be used to date such objects and worked out the necessary techniques. As a result, he was awarded the Nobel prize in chemistry in 1960.

Radiocarbon dating isn't easy. If you take a quantity of present-day wood, you will get only thirteen low-energy beta particles per minute for every gram of carbon it contains. If it is five thousand years old, you will get perhaps seven counts per minute. These beta particles have to be detected despite the various other radiations in the environment that are *not* produced by carbon-14. This means that the counting device has to be surrounded by elaborate shields.

The accuracy of the technique can be determined by working out the age of wood from old Egyptian tombs and comparing it with the age as determined from historical evidence. The result isn't bad, though radiocarbon dating doesn't pin things down as precisely as the historic evidence seems to.

You might think that if ordinary historical reasoning is more precise than radiocarbon dating, we have no need for the latter—but Egyptian relics take us back only five thousand years. Before that, there stretches a period of prehistory for which ordinary dating is very vague indeed, but for which radiocarbon dating remains reasonably accurate. Radiocarbon dating can, indeed, work for objects up to seventy thousand years old.

Radiocarbon dating has been used to give us an idea of when human beings first entered the Americas, for instance, and when the most recent retreat of the glaciers took place. As a matter of fact, it had been reasoned out that the last retreat of the glaciers took place perhaps twenty-five thousand years ago, but radiocarbon data for samples of ancient wood tell us this happened only ten thousand years ago.

Can we be sure that radiocarbon dating is accurate, however? Are there any sources of error?

Assuming that the decay rate of carbon-14 is constant over the eons (as physicists are confident it is), then one source of error is fractionation. Carbon-14 is about 4.5 percent more massive than carbon-12, and while the former undergoes the same chemical reactions as carbon-12 does, it does so a bit more sluggishly. That means that if you start off with a quantity of carbon and allow half of it to react in a certain way, the portion that has reacted is richer in carbon-12 and poorer in carbon-14 than the portion that has not reacted. Such fractionation effects must be taken into account—and are.

A more troubling kind of error involves the formation of carbon-14 to begin with. After all, how can we assume that the incidence of cosmic-ray particles has always been constant? Might not the number of particles striking the atmosphere vary over the years?

Periodically, a supernova may explode within a few hundred light-years of Earth. Wouldn't that mean that there would be a temporary wash of additional cosmic-ray particles over the Earth?

Then, too, variations in the intensity of Earth's magnetic field would result in cosmic-ray particles being warded off with varying efficiency, and we know that the magnetic intensity *does* vary considerably over the years.

Some idea of the variations in cosmic-ray intensity and the rate of formation of carbon-14 can be obtained by studying the carbon-14 content of various rings in old wood, and this helps us make allowance for such variations.

But cosmic events like supernovas and planetary magnetic-field variations are not all that introduce uncertainties. Believe it or not, human activity now does the same. For a couple of decades after World War II there was atmospheric testing of nuclear bombs and, as a result, large numbers of neutrons were released into the atmosphere. This resulted in the formation of enough carbon-14 to raise the total quantity significantly.—And thereby hangs a tale.

Suppose we ask ourselves to what extent the human body is affected by radioactivity in the natural environment. There are small quantities of uranium and thorium in the rocks and soil that surround us; in the bricks and stones of which houses are built; and so on. In fact, uranium and thorium, breaking down, produce infra-tiny quantities of a radioactive gas, radon, and these days people are worried about the accumulation of radon in the air inside houses. This is particularly so since we're now so busy insulating, in order to conserve heat, that we are also cutting down on the ventilation that would sweep radon out of our living quarters and into the atmosphere.

In addition, there is the unending pitter-patter of cosmic-ray particles and the secondary radiations they produce, which penetrate our bodies constantly throughout life.

All this energetic radiation can disrupt molecules within our bodies, occasionally producing mutations which can show up most drastically in the form of occasional development of cancer and birth defects.

However, humanity (and all life) has been subject to this sort of thing throughout its history and the destructive effects of such external radiation are less than the constructive effects—since a certain level of mutation is necessary if evolution is to take place at some reasonable rate. Without the radiation that can occasionally produce a fatal cancer or birth defect, we wouldn't be here at all—so the price must be paid.

Besides, external radiation is not as bad as it sounds. As it penetrates and passes through our bodies, such radiation has very little chance of striking any molecule whose disruption will cause a mutation. Most of the time—by far, most of the time—such radiation expends itself on water molecules and on other relatively insensitive constituents of the body.

But not all radiation is external. The body is itself radioactive. There is an enemy within!

The body is composed of various elements and some of them have naturally occurring radioisotopes. One of the elements is potassium, an absolutely essential component of the body. In nature (and in our bodies) there are three potassium isotopes, potassium-39, potassium-40, and potassium-41. Of these, potassium-40 is the rarest. Only 1 potassium atom out of 8,400 is potassium-40. This potassium-40, however, is very slightly radioactive. It has a half-life of 1.3 billion years and is therefore constantly producing beta particles.

The human body is about 0.01 potassium. A 70-kilogram (154 pound) adult would therefore contain 700 grams (24.7 ounces) of potassium. This means it contains 83 milligrams (three thousandths of an ounce) of potassium-40. We can figure out how many atoms of potassium-40 are present in 53 milligrams, and from the half-life, we can figure out how many of these atoms are breaking down and releasing beta particles each second. The answer is 1,900 a second.

These beta particles disrupt atoms and molecules and do damage. However, there are 50 trillion cells in the body and, on the average, any one cell is exposed to the effect of only one potassium-40 beta particle per year. And, again, this beta particle usually expends its energy in harmless ways.

It has been calculated, in fact, that potassium-40 radiation subjects the body to radiation of about the same order of magnitude that cosmic rays subject it to. Since we can live with cosmic rays, we can also live with potassium-40.

No other element essential to the body's functioning has a natural long-lived radioactive isotope. Two of the elements, however, have a short-lived radioactive isotope that wouldn't exist but for the fact that it is constantly being manufactured by cosmic rays. One of them is carbon-14, of course, and the other is hydrogen-3 (tritium).

In 1946, Libby showed that cosmic rays form hydrogen-3, which is thus found in nature in small quantities. Hydrogen-3 has a half-life of 12.26 years, which is only $\frac{1}{460}$ that of carbon-14. It vanishes correspondingly more quickly, so that the concentration in the atmosphere (and therefore in plants, and therefore in us) is much lower than is the case with carbon-14.

In naturally occurring hydrogen, only 1 atom out of a billion billion is hydrogen-3. The human body is about 0.12 hydrogen, but this includes only 8.4 quadrillionths of a gram of hydrogen-3, a vanishingly small quantity. Hydrogen-3 gives rise to only three disintegrations per second in the body as a whole. This can be dismissed as completely insignificant.

That leaves us with carbon-14. The human body is 0.15 carbon, so a 70-kilogram person contains 10.5 kilograms of carbon. Since there is 1 carbon-14 atom in every 540 billion carbon atoms, the body contains 190 millionths of a gram of carbon-14. Given the half-life of carbon-14, we can calculate that the number of beta particles produced by carbon-14 in one second is about 3,100.

This means that the total number of beta particles produced in a 70-kilogram human body is about 22,100 per second. Of these, 86 percent is produced by potassium-40, 14 percent by carbon-14, and 0.00014 percent by hydrogen-3.

Since I reasoned that the potassium-40 in the body is not present in great enough quantity to be considered any more dangerous than cosmic-ray bombardment, it might seem we could certainly dismiss carbon-14 and hydrogen-3 and drop the whole subject.

But wait! Let's start all over again.

The various parts of the body are not all to be considered equally vital. We know this. A bullet in the shoulder or the foot is no pleasure, but it probably won't kill you. A bullet in the brain or heart, however, will finish you at once.

In the same way, an energetic particle streaking through a cell may hit a number of water molecules, or fat molecules, or starch molecules and won't do any irreparable damage. That same particle hitting a DNA molecule can do much damage, for the DNA molecule controls some vital portion of the cell machinery and damage done to it can produce a mutation that *may* bring about cancer or birth defects.

However, the mass of the DNA molecules in the cells is about 1/400 of that of the entire cell, so that particles streaking through the cell in random directions will not often strike a DNA molecule and will expend their energies (as a bullet in the shoulder will) on relatively unimportant changes. This is true even if the particle is produced by some breakdown within the body.

In other words, most radiation originating within the body is not very different in its effect from that of radiation that comes from outside the body. It is only if the radioactive atom happens to be actually located within the DNA molecule itself that we have a true case of the enemy within.

From that standpoint, potassium-40 is eliminated. There are no potassium atoms in the DNA molecule. There are, however, carbon and hydrogen atoms present, so there must also be present, to some small extent, carbon-14 and hydrogen-3.

Of these two, the hydrogen-3 atom produces only one thousandth as many breakdowns as the considerably more common carbon-14 atom, so let's eliminate hydrogen-3 as probably insignificant and concentrate on carbon-14.

Every time a carbon-14 atom breaks down, it becomes a nitrogen-14 atom. This change of carbon to nitrogen changes the chemical nature of the DNA molecule and this, in itself, produces a mutation, though how dangerous a mutation, it is hard to say. However, when the carbon-14 atom shoots out a beta particle, the chemical change may be the least of it. There is a recoil which may force the exploding carbon-14 to break the bonds that hold it to its neighbors. In other words, the DNA molecule will break in two and this makes for a possibly drastic kind of mutation.

Suppose we calculate how many carbon atoms there are in the DNA molecules of a cell and then how many of these are carbon-14. I've done this in a sort of back-of-the-envelope way and it seems to me that there is 1 atom of carbon-14 for every 20 cells, and 1 breakdown per year for every 24,000 cells.

That doesn't sound like much, but, again, there are about 50 trillion cells in the body, so that we end by having about six breakdowns of carbon-14 in all the various DNA molecules of a 70-kilogram body *every second.*

What's six breakdowns per second? Virtually nothing, you might suppose, and if they were ordinary breakdowns with particles speeding through the cell at random, you would be right. In this case, however, *every single one of the breakdowns produces a mutation at the very moment of breakdown.*

It is possible, of course, that most of these mutations are relatively harmless. It is also possible that some severe mutations may kill a cell which may then be easily replaced.

However, some cells killed in this way (notably nerve cells and brain cells) may not be capable of replacement. Also some mutations may not kill a cell but may make that cell a cancerous one. It might be argued that the important mutations found in all organisms result principally (though, of course, not entirely) from the carbon-14 atoms present in DNA molecules, and that the effect of cosmic rays, for instance, rests indirectly upon the carbon-14 atoms they form.

I first pointed out the danger of carbon-14 in DNA molecules in a short article entitled "The Radioactivity of the Human Body" in the February 1955 issue of *The Journal of Chemical Education.* (Yes, for a few years in the early 1950s I wrote articles for learned journals. This one, as it happens, was the last.)

I believe I may have been the first, or very nearly the first, to do so. Willard Libby may have beaten me to the punch by a few months, but I'm not sure of that. In any case, I wasn't aware of his work when I wrote my article.

The last paragraph of my paper went as follows: "In the light of this, it would be interesting to note whether a diet high in carbon-14 would increase the mutation rate in an animal such as *Drosophila,* or the rate of tumor formation in cancer-prone strains of mice, and whether any correlation existed between the increase (if any) in mutagenesis or carcinogenesis, and the increase (if any) of carbon-14 in the genes."

I don't know if such experiments were ever carried out. Certainly, I didn't have either the training or the equipment to carry them out myself. I also didn't know at the time I wrote the article that the atmospheric testing of nuclear bombs was producing a significant increase in the atmospheric content of carbon-14.

Linus Pauling, however, did know of the increase, and, sometime later, he saw its significance (and I can only hope that my article in *The Journal of Chemical Education*—a journal he later told me he read regularly—contributed to the realization). He promptly began a campaign to convince world leaders and the public that every nuclear explosion in the atmosphere increased the incidence of cancer of various sorts and of birth defects because of the increase of carbon-14 in the atmosphere and, therefore, in the genes.

It was his arguments of this sort, more than anything else, that led to the test-ban treaty of 1963 and to the end of atmospheric nuclear explosions.

I'm rather proud of this. My own role was microscopic and I give all credit to Professor Pauling, but of all the good scientific ideas I have had in my lifetime, and I have had a few, I think this one was the best.

5

The Light-Bringer

Yesterday, I was interviewed by a Soviet newsman before the television camera. I am interviewed by the Soviets now and then, you see, because my science fiction is popular in the Soviet Union, and because I was born there.

Usually, I am interviewed on the issues of peace, love, and cooperation among nations, and I always assure them I am in favor of all three, and I usually wax eloquent on the subject. Yesterday, however, I was interviewed on the subject of science fiction and of myself and, as you can well imagine, having this favorite of all subjects of mine brought up caused my eyes to blaze with supernal light and my eloquence to reach incredible heights.

When it was time for me to discourse on the subject of robotics, I stopped suddenly, and said, "I invented the word, you know." The interviewer registered interest and surprise and I went into full details.

I thought about this afterward. Such is my intense

devotion to claiming credit that is mine that I have managed to get my invention of the word mentioned in American dictionaries—and now I am spreading the glad tidings over the length and breadth of the Soviet Union as well. But is this fair?

Think of all the great discoverers who are forever lost because they didn't have modern communications at their disposal. Someone must have invented the wheel, but how could he broadcast or preserve the great news of what he had done?

No one knows who first tamed fire, who first caught the trick of melting copper out of blue rock, who first got the idea of tying up goats and stealing their milk, or who first said, "Hey, let's plant grain, take care of it all summer, and then have a lot of food in the winter." It seems a shame, in view of this, that I should be in a position to force the whole world to remember that I made up a word.

Of course, out of a series of similar discoveries there must come a time when the name of one actual discoverer comes to be remembered. For instance, who was the first person, known by name, who discovered a chemical element? What was the element and when was it discovered? As usual, I shall start at the beginning.

Of the hundred-plus elements now known, at least nine were known even in early ancient times. They were not recognized as elements (the various fundamental substances making up the Universe at the atomic level) at the time, for the ancients had their own, mistaken, notions as to what elements were, but never mind that. We'll talk about elements in accordance with contemporary notions.

Seven of the early known elements were metals. These were known because they happened to exist, in small quantities, in reasonably pure elementary form, and because that elementary form was easily recognizable.

Thus, if someone happened to come across a gold nugget, he or she would be immediately aware of something yellow and shiny that looked quite different from ordinary pebbles. In addition to appearance, it would be much heavier than other pebbles of similar size and, when struck with a stone ax, would neither flake nor shatter, but would deform. Given its beauty and workability, it is not surprising that gold ornaments are found in prehistoric graves in Egypt and Mesopotamia.

Because of its properties, gold was sought after, and because it is one of the rarest of the elements, discoveries were few, and therefore notable. Other similar substances were also sought. The very word "metal" comes from the Greek word meaning "to seek."

Silver is perhaps twenty times as common as gold, but it is also more chemically active, and, therefore, more likely to exist in combination with other elements as "ores." These ores lack metallic properties and look much like ordinary rocks. Silver nuggets were therefore discovered later than gold nuggets were, but were still known in prehistoric times.

Later, when it was learned how to separate a metal from its ores by heating the latter under the proper conditions, silver became more common than gold.

Copper is perhaps 450 times more common than silver and 9,000 times more common than gold, and even though it is more chemically active than either of the other two, copper can be found, not too rarely, in the elementary state. It is possible that copper was used for ornaments even earlier than gold was. Once the smelting of copper ore was worked out, copper could even be used in massive quantities as a component part of tools and weapons.

Iron is one of the most common of the elements, over a thousand times more common than copper, but it is so active that under ordinary conditions it is always found as ore and not in the elementary form at all. It is also much more difficult to smelt iron ore than to smelt silver or copper ores. In fact, it was not until 1500 B.C. that the Hittites worked out a practical method of smelting iron ore.

Nevertheless, metallic iron falls from the sky in the form of an occasional meteorite and, thanks to these meteorites, iron was known in its elementary form, even in prehistoric times.

Lead is only a third as common as copper, but it is easily obtained from its ore. When people were smelting ores to get the desirable silver and copper, any lead ores that happened to be thrown in would yield lead.

Lead was as dull and ugly as gold was lustrous and beautiful, so that where gold was the "noble metal" par excellence, lead was the very epitome of the "base metal." Nevertheless, lead had its value. For one thing, it was the densest substance known to the ancients, except for gold, so that if one wanted an object to be both small and heavy, and if one could not afford gold, one used lead as next best. For another, lead was quite soft and could easily be molded into pipes through which water could be led. These eventually replaced clay pipes, which were too easily broken, so that the word "plumber" comes from the Latin word for "lead."

Tin was probably discovered indirectly. Copper ores that yielded relatively pure copper produced a metal that was too soft to use for tools, weapons, and armor. But if, to the copper ore, another ore was added, a metal alloy was produced which was a great deal harder than copper itself. This mixture is called "bronze" and the mystery additive is the metal tin. The heroes of the Trojan War had bronze shields, bronze armor, and bronze spear-points. They lived in the "Bronze Age," which succeeded the "Stone Age" and was itself to be succeeded by the "Iron Age."

Tin could be smelted out of its ores and then combined with copper in proportions that best balanced quality and price. However, tin is only about a fifteenth as common as copper, and the tin mines of the Mediterranean region were played out rather early on. (This was the first disappearance of a vital resource in history.) The Phoenicians then ventured out into the Atlantic to locate tin ore in the "Tin Islands" (usually equated with Cornwall) and make themselves rich in consequence.

Mercury was the last of the ancient metals to be discovered and was, of course, very notable for being a liquid.

In addition to these seven metals, there are two nonmetals that occur, very noticeably, in the elementary state. One of these is sulfur, which is a pronounced yellow in color, but totally without the beautiful metallic luster of gold. People couldn't have helped but come across it in ancient times.

The most noticeable thing about sulfur was that it burned, as people must inevitably have noticed if they tried to build a campfire in the vicinity of some sulfur. All the common fuels known to the ancients were derived from living things: wood, oils, and so on. Sulfur was the only substance with no connection with life that burned readily, so it was called the equivalent of "burnstone," for instance, which, in English, was corrupted to "brimstone."

The burning of sulfur is very noticeable because not only does it burn with an eerie blue flame, but it also releases an unendurably irritating gas in doing so. That, combined with the noticeable presence of this irritating odor in the neighborhood of active volcanoes, undoubtedly gave rise to the notion of an underground hell in which there was not only unending fire, but the additional unpleasantness of sulfur as a major fuel (hence, "fire and brimstone").

Finally, there is carbon. Any campfire built near a rock or inside a cave is going to leave a deposit of soot on the rock, and this soot is virtually pure carbon. Again, if a pile of wood is burned under conditions where there is limited access to air, the wood in the interior of the pile does not burn completely. A black substance remains there, which, if ignited under conditions where plenty of air will reach it, will burn with less flame and with considerably hotter temperature than the original wood would have. The black substance is charcoal and, again, it is virtually pure carbon.

Clearly, ancient man must have been aware of the existence of soot and of charcoal.

* * *

In addition to these nine elements, there are several more that must have been isolated at least by medieval times, but about whose early history we know very little.

For instance, before the ordinary copper-tin mixture we call bronze came into use, the early copper workers had found that copper ore mixed with another kind of ore (*not* tin ore) also produced a copper alloy that was considerably harder than pure copper.

The trouble was that working with this earlier bronze was dangerous and the death rate among those who dug up the other ore and mixed it with copper ore was high. As it happened, the other ore was an arsenic ore, and when tin ore came in, arsenic ore, very sensibly, went out.

Of course, discovering and using an ore is not the same thing as isolating the element it contains. However, once human beings learned to get such metals as copper, tin, lead, mercury, and iron out of their respective ores, it seems sensible to suppose that any ore in which an element is not particularly tightly fixed would be successfully smelted.

From arsenic ore, it is not difficult to obtain arsenic itself and it must have been done in ancient and early medieval times on a number of occasions. In those days, however, scientific discoveries were not particularly publicized, if no useful applications were involved. The arsenic ores were poisonous and few people must have worked with them. Any arsenic obtained from them had no particular use, and was forgotten.

The first person to force elementary arsenic into the

consciousness of the scholarly world was Albertus Magnus (1193–1280), a German scholar. He prepared it from its ore and described it in his writings carefully and accurately enough to leave us in no doubt that it was arsenic he obtained. For this reason, Albertus Magnus is sometimes considered to have "discovered" arsenic about 1230. If that were so, he would be the first person, acknowledged by name, date, and place, to have discovered an element, but that is not strictly legitimate. There is every likelihood that arsenic had been isolated much earlier by people whose names are not known.

Then, too, there are black pigments that were used in ancient time to darken the eyebrows and eyelids, much as moderns use mascara. It may have been used in Egypt as long ago as 3000 B.C. One of the pigments so used was called "stibium" by the Romans, and "stibnite" in modern times. The pigment is, chemically, antimony sulfide.

Antimony is similar to arsenic in its chemical properties and since the latter can easily be extracted from its sulfide ore in elementary form, so can the former. What's more, it has been. There is a vase obtained from an ancient Mesopotamian site, possibly dating back to 3000 B.C., that is almost pure antimony. Other ancient relics containing antimony have also been found.

The first scholarly discussion of antimony was in a book entitled *Triumphal Chariot of Antimony.* It was supposed to have been written in 1450 and was attributed to a German monk named Basil Valentine. For that reason, Valentine is sometimes listed as the discoverer of antimony, but, of course, he was not. In fact, there is no clear evidence that he ever lived and the book itself may have been written about 1600 by someone who ascribed it to a monk of the past in order that it might be taken more seriously.

The element bismuth, which is also a member of the arsenic family of elements, may have first been isolated in the 1400s, or, some think, even earlier. Its discovery is also sometimes attributed to Valentine, but we can be certain that the real discoverer is unknown, and the actual time of discovery earlier.

Finally, there is zinc. In ancient times, zinc ores were mixed with copper ores and the resulting copper-zinc alloy was "brass." The distinction of brass is that it has a color that very closely resembles that of gold. It has none of the other properties of gold, but there are times when merely a similarity of appearance is enough.

It would have been very easy to obtain elementary zinc from its ores, except that at the high temperature of smelting there is a tendency for zinc to vaporize and disappear. (Zinc is a member of the family of elements to which the low-melting, low-boiling mercury belongs.) Nevertheless, it is quite likely that elementary zinc was obtained in Roman times.

This, then, was the situation in 1674. Some thirteen substances, now recognized as elements, were known at the time. In alphabetical order, they are: antimony, arsenic, bismuth, carbon, copper, gold, iron, lead, mercury, silver, sulfur, tin, and zinc. All were known in reasonably pure form but the discovery of not one of them can truly be pinned down to a particular time, place, or person.

And all this brings us to phosphorus.

* * *

The word "phosphorus" entered the scientific vocabulary in ancient times. A very bright star appears sometimes in the western sky after sunset, while a similar one appears at other times in the eastern sky before dawn. They were the "evening star" and the "morning star," respectively. At first the Greeks considered them as two separate objects. They called the evening star "Hesperos" (or "Hesperus" in Latin and English spelling) from their word for "west," and they called the morning star "Phosphoros" (or Phosphorus in Latin and English spelling) from their words for "light-bringer." The reason for the latter name was that once the morning star rose in the east, one could be certain that the dawn would soon come.

The Romans gave the two objects the Latin names with the same meaning as the Greek names—"Vesper" for the evening star and "Lucifer" for the morning star.

Eventually, though, the evening star and morning star were recognized as the same object (thanks to the more advanced Babylonian astronomy, undoubtedly) and the two names fell out of use. The star (or planet, actually) came to be known as "Aphrodite" to the Greeks and as "Venus" to the Romans and to us.

And with that "phosphorus" disappeared from the scientific vocabulary for a little over two thousand years, until we reach the time of Hennig Brand, a German chemist, who was born about 1630 and who died about 1692.

Brand worked in the tradition of the alchemists (he is sometimes called "the last of the alchemists") and was interested in discovering some substance that would catalyze the conversion of base metals to gold or, at the very least, the conversion of silver to gold.

It fell into his head (we don't know why) that he might obtain such a catalytic substance from urine. In 1674, he therefore engaged himself in the rather smelly business of boiling down a large quantity of urine until he had isolated the dissolved material as a solid crust in his vessels. This contained, among other things, what we would call "sodium phosphate."

He then treated the solid residue in the usual way in which ores were smelted to see if he could extract a new metal that would serve as a catalyst for the production of gold. When so treated, the sodium phosphate gave up some of its phosphorus atoms and Brand was able to isolate some reasonably pure phosphorus.

No one had ever seen elementary phosphorus before; no one had ever suspected its existence. It was the first element to be isolated in modern times, the first to be isolated at a known time (1674), in a known place (Hamburg, Germany), and by a known person (Hennig Brand).

But why get excited over this? Of course, discovering a new substance with properties like nothing earlier known is exciting, but there was more to it than that.

The thing is that the new substance glowed greenly in the dark. That was a mysterious and eerie property and Brand named his discovery "phosphorus," for it was a "light-bringer"—and thus the word reentered the scientific vocabulary in a manner totally different from that used by the ancient Greeks.

To be sure, there were minerals that glowed in the dark, a phenomenon now called "phosphorescence" (which has nothing particularly to do with phosphorus, despite the similarity in names). Phosphorescence, however, takes place only after the mineral has been exposed to the light, and the light it produces in darkness fades rather quickly as time passes. Phosphorus, on the other hand, glows even when it has not been exposed to light, and the glow continues for a long time.

The glow raised the same excitement among chemists of Brand's time as the glowing radium that Marie Curie isolated did among the chemists of more than two centuries later. (There is a difference, of course. Phosphorus glows because it spontaneously and slowly combines with oxygen, releasing chemical energy that is converted, in part, into light. Radium glows because its nucleus breaks down spontaneously, producing nuclear energy which is converted, in part, into light.)

Thanks to the excitement stirred by the glow, other chemists tried to obtain phosphorus for themselves. One came to Brand for instructions, and, having received them, proceeded to make phosphorus and then claimed (unsuccessfully) to be the actual discoverer. The British chemist Robert Boyle (1627–91) actually isolated phosphorus independently in 1680, but he was too late by six years and Brand gets the credit.

Phosphorus belongs to the same family of elements to which arsenic, antimony, and bismuth belong. Antimony and bismuth are metals and arsenic is characterized as a semimetal, but phosphorus, made up of atoms distinctly smaller than those of the other three, is definitely not a metal. As Brand prepared it, it is a white, waxy solid, so that it is frequently called "white phosphorus."

Naturally, one looked for a way of putting this connection of phosphorus and light to use. The German scholar Gottfried Wilhelm Leibniz (1646–1716) enthusiastically suggested that a large enough piece of white phosphorus could be used to light a room, thus making candles unnecessary.

However, the difficulty of manufacturing phosphorus is such that a slab of material large enough to light a room would cost enough to keep someone in candles indefinitely.

But then, glowing phosphorus gave off heat as well as light and if it came into contact with something inflammable it might, after a while, set it on fire. In fact, chemists, who were careless with phosphorus at first (as later chemists were careless with radium at first) did succeed in inadvertently setting fires in their homes and workplaces.

That raised the question of starting a fire by chemical means.

Until then, fires had been started by the use of friction. One piece of wood would be ground into another until there was enough heat developed to ignite some tinder, and the small fire could be used to light a bigger one. Or else, flint and steel could be struck together to create a spark of burning iron that would ignite tinder.

But why not simply coat the edge of a splint of wood (or of heavy paper) with some appropriate chemical that could, at the proper time, set the wood or paper on fire? You would then have a small fire that would last long enough to ignite a larger and longer-lasting one. In short, you would have a match (from an old word for the nozzle of a lamp, where the burning oil produced a flame, to which the flame produced by the wooden splint was likened).

Such chemical matches began to be produced in the early decades of the nineteenth century. Some did not make use of phosphorus. One type had a moist mixture including the active chemical potassium chlorate enclosed in a glass bead at the end of a stick, the whole being wrapped in paper. When the head was broken, the potassium chlorate set the paper on fire. These were called "Promethean matches," from Prometheus, the god in the Greek myths who had brought fire from the Sun to human beings. This was a very slow and messy kind of match, you could well believe.

Another kind of match did not catch fire spontaneously. You had to increase the temperature by striking it; that is, by rubbing it over a rough surface. The friction developed heat which caused the active tip to undergo a chemical change that caused it to burst into flame. Such "friction matches," made without phosphorus, were called "Lucifer matches" from the Latin word for light-bringer.

Such matches have a minor role in American history. The Americans called them "loco-foco matches," partly because "loco-foco" seemed to mean "self-lighting" by analogy with "locomotive," which could be looked on as meaning "self-moving," and partly by distortion from "Lucifer."

In 1835, the liberal wing of the Democratic Party in New York City was in a hot dispute with the conservative wing. At a party meeting, the conservatives, scenting defeat, put out the lights to end the meeting. The liberals, however, lit candles with their loco-focos and continued. After that, for a while, the conservative

Democrats called the liberals "Loco-focos" as an expression of contempt and the other party, the Whigs, gladly applied the name to all Democrats.

The Lucifer matches, without phosphorus, were hard to strike and when they finally did catch, they sometimes emitted a shower of sparks that could produce burns to clothes and hands.

In 1831, however, a Frenchman named Charles Sauria produced the first practical friction match containing phosphorus, diluting the active phosphorus with other materials to make sure that the matches didn't start to flame until they were struck. Such matches produced flame quickly and quietly when struck and didn't deteriorate on standing. They eventually put all other varieties of matches out of business.

There was one catch. The phosphorus used in the matches was quite poisonous and people who worked at producing the matches would get the phosphorus into their bodies, where it caused bone degeneration. They got what was called "phossy jaw" and it killed them, slowly and painfully.

Here, too, there was a peculiar analogy with what happened a century later with radium. The danger of radium and radioactive substances was not appreciated at first, and radium was incorporated in tiny quantities into material that was painted on clock and watch faces to make the numbers and hands glow in the dark.

Those who worked in factories with the radium got radiation sickness and died, and the whole thing was finally outlawed. (I remember wearing a radium-painted watch when I was young.)

Fortunately, in 1845, an Austrian chemist, Anton von Schroetter (1802–75) discovered that if white phosphorus was heated in an atmosphere of nitrogen or carbon dioxide (with which it won't react), its atoms rearrange and it becomes another kind of phosphorus called, from its color, "red phosphorus."

The advantage of red phosphorus is that it is nontoxic and can be used with relative safety; so by 1851, Schroetter was producing and recommending red phosphorus matches. However, red phosphorus is not as active as white phosphorus so that a red phosphorus match is not as easy to strike into flame. For this reason, white phosphorus matches continued to be popular till the end of the century, when they were outlawed. Society, forced to choose between systematic death and a little inconvenience, chose, with its usual delay and reluctance, the inconvenience. Eventually, though, the red phosphorus matchheads were chemically juiced up to the point where they were perfectly easy to strike.

The next step was to produce "safety matches." Ordinary matches could be struck on any rough surface since all the chemicals necessary to produce a chemical action leading to heat and flame were in the matchhead. Accidental lighting leading to unintended destruction, injuries, and death might take place.

Suppose, though, that you left out one ingredient from the matchhead—the red phosphorus, for instance—and placed it on a special strip. The safety match, containing various chemicals, but no phosphorus, will then ignite only if it is struck on the strip.

But that's enough for now. I will have more to say about phosphorus in the next chapter.

6

Beginning with Bone

The other day I found myself trapped on a dais at a luncheon at which I was not scheduled to talk. That, in itself, placed a frown on my youthful face. Why chivvy me into sitting at the dais instead of at a table with my dear wife, Janet, if they were not going to be making any use of me.

Of course, I would be introduced, which meant I could at least rise and smile prettily. It turned out, though, that the introducer had never heard of me, and so mangled my name when she tried to pronounce it that I aborted my rise quickly and refused to smile.

So it didn't look as though it was going to be my day. In sheer desperation, I occupied myself in writing a naughty limerick for the woman who sat to my left, and who *did* know me. (In fact, she was responsible for my being stuck there.)

I suppose she noted that I was looking rather grim, so she undertook to cheer me up by bringing me to the attention of others. She turned to the man on *her* left and said, "Look at this funny limerick that Dr. Isaac Asimov has written for me."

The businessman looked at it with lackluster eye and then looking up at me, said, "Are you a writer, by any chance?"

Well! I don't expect people to read my stuff necessarily, but read it or not, I do expect them to have at least a vague suspicion that I'm a writer.

My friend at the left, noting that my hand was creeping toward the knife beside my plate, said hurriedly, "Oh, he *is.* He has written three hundred fifty books."

Unimpressed, the fellow said, "Three hundred fifty limericks?"

"No, three hundred fifty *books.*"

There then took place the following conversation between the man and me.

MAN *(unwilling to let go of the limericks):* "Are you Irish?"

ASIMOV: "No."

MAN: "Then how can you write limericks?"

ASIMOV: "I was born in Russia and write odessas."*

MAN *(looking blank:)* "Do you use a word processor?"

ASIMOV: "Yes."

MAN: "Can you imagine getting along without one now?"

*In the very unlikely chance that you don't get this, Limerick is a town in southwestern Ireland, and Odessa one in southwestern Russia.

ASIMOV: "Sure."

MAN *(paying no attention):* "Can you imagine what might have happened to *War and Peace* if Dostoevski had had a word processor?"

ASIMOV *(scornfully):* "Nothing at all, since *War and Peace* was written by Tolstoy."

That ended the conversation and I turned my attention to surviving the lunch—which I did, but not by much.

All is not lost, however. Having met with a bonehead, it struck me that I would start my next *F & SF* essay by considering bone.

Life, as we know it on Earth, is carried on in a watery base in which molecules of various sizes are dissolved or suspended. On the whole, this means that life-forms are apt to be soft and squashy, like earthworms, for instance. It is possible to get by in a soft and squashy way, and all of life managed to do so until the Earth was nearly seven-eighths its present age. It is only comparatively recently that life developed hardness.

Of course, even at its softest and squashiest, bits of life couldn't simply exist as watery solutions immersed in the ocean. They would be dispersed and washed away. Each bit of life had to have some outer pellicle that would keep the molecular machinery of life together, and separate it from the surrounding ocean.

This was done by building up macromolecules (chains of small molecules) to form cell membranes. Plant cells, concentrating on sugar units, built up cellulose out of long chains of glucose molecules, and that is now the most common organic molecule in existence. Cellulose is the major component of wood. Cotton, linen, and paper are practically pure cellulose.

Animal cells do not make cellulose. They concentrated on other macromolecules (proteins, for instance) for doing the job of coherence. The tough protein keratin is a major component of skin, scales, hair, nails, hooves, and claws. Another tough protein, collagen, is to be found in ligaments, tendons, and in connective tissue generally.

But about 600 million years ago, quite suddenly on the evolutionary scale, various animal groups ("phyla") developed the trick of using inorganic substances as protective walls. These were essentially rocky in nature, and were harder, stronger, and more impervious to the environment than anything built out of organic materials. (They were also heavier, less sensitive, less responsive, and often forced those creatures weighed down by the material to take up a motionless life.)

These "skeletons" served not only as protection, but as a good place to attach muscles, which could then pull harder and more powerfully. Furthermore, it is these hard parts that make up the bulk of the fossil remnants of life that we find in sedimentary rock. Being rocklike in nature, they can easily undergo changes (under the proper circumstances) that make them more rocklike still. They can then retain their original shape and form for hundreds of millions of years. It is for this reason that fossils are common only in rocks younger than 600 million years. Before that, there were no hard parts to fossilize.

The simplest animals to develop a skeleton were the one-celled "radiolarians." These microscopic creatures have beautiful skeletons of intricate inorganic spicules composed of "silica" ("silicon dioxide")—which is the characteristic substance making up sand.

Silica, however, although exceedingly common, did not become the general skeletal material. It is apparently too difficult for organisms to handle. Human beings, for instance, in common with animal life generally, do not contain any silicon compounds as essential parts of our bodies. Any such compounds present are just temporarily there as impurities swallowed with our food.

Beginning with the simplest multicellular animals, there developed a tendency for forming skeletons made out of calcium compounds, particularly calcium carbonate, which is also known as "limestone."

The seashells of members of the phylum Mollusca (clams, oysters, snails) are made of calcium carbonate. This is also true for members of other phyla such as coral, bryozoans, lampshells, and so on. For that matter, the eggshells formed by reptiles and birds are also calcium carbonate.

The phylum Arthropoda, however, struck a compromise. They did not get bogged down beneath a heavy shell that left them with oysterlike immobility. They avoided inorganic strengthening altogether and remained with organic macromolecules. They improved on those, however.

The arthropods (which include such creatures as lobsters, crabs, shrimps, insects of all kinds, spiders, scorpions, centipedes, and so on) all have a skeleton of "chitin," from a Greek word for a coat of armor, or shell.

Chitin is a macromolecule built up of sugar units very much as cellulose is, but with a difference. Whereas cellulose is built of glucose units (glucose being a very common and simple sugar), chitin is built of glucosamine units. The glucoses in the chitin chain are each modified by the presence of a small nitrogen-containing group, and this suffices to make chitin quite different from cellulose in its properties.

Chitin is tough enough to serve as protection, it is flexible as well, and it is light enough to allow rapid, active movement. Indeed, insects, despite their thin skeletons of chitin, are mobile enough to fly. (Of course, they can do so only at the cost of remaining very small.)

Chitin may well be one of the reasons why the arthropods are so amazingly successful. There are far more species of arthropods than there are species of all the other phyla put together.

This brings us to Chordata, the last phylum to come into existence (from starfishlike ancestors) about 550 million years ago. What differentiates chordates from all other creatures is that they have, first, a nerve cord that is hollow and not solid, and that runs along the back and not along the belly. Second, they have gill slits through which they can pass water and filter out food (though in land-dwelling chordates, these show signs of developing only in the embryonic stage). Third, they have a stiffening rod, called a notochord, running parallel to the nerve cord (though, again, the notochord may only be present in embryonic or in larval stages).

The notochord is made up of collagen, chiefly, and is an example of an internal skeleton, rather than the external skeletons found in other phyla. Internal skeletons are found, to a fumbling extent, in a few other instances in the other phyla, but only the chordates went on to specialize in it. They went further than the arthropods. They left the outer skin unprotected, and left the skeleton inside where it could serve to maintain shape and integrity, and be an anchor to muscles. The softness and vulnerability of unprotected skin is more than made up for by the strength, power, and mobility that chordates can develop, thanks to the relatively light but strong internal skeleton. It is no wonder that the largest, most powerful, fastest, most intelligent, and, in general, the most successful animals who have ever lived are chordates.

Early in chordate history, the simple rod of the notochord was replaced by a series of separate bits of skeletal tissue that actually enclosed the nerve cord, giving it additional protection. These separate bits of skeleton are called "vertebrae" and they make up the "spinal column." Nowadays, all chordates, but for three groups of very primitive out-of-the-way organisms, possess vertebrae. They make up the subphylum, Vertebrata, and are the "vertebrates."

The earliest vertebrates, which evolved about 510 million years ago, were the first to develop bone. This was an inorganic skeletal material that was composed of calcium compounds, but was not quite calcium carbonate. This bone was restricted to the outside of the body, especially in the head region and these early vertebrates were called "ostracoderms" (from Greek terms meaning "shellskin"). The vertebrae inside the body were cartilage, which was chiefly made up, again, of collagen.

The external skeleton of the ostracoderms limited mobility, however, and, in general, this was not a successful device. Vertebrates without outer armor, who relied on mobility and agility, did better. Even today those chordates with outer armor, such as turtles, armadillos, and pangolins, are not notably successful.

Ostracoderms developed in two directions. They developed more elaborate internal skeletons, including cartilaginous extensions that made four limbs possible, and other extensions that made movable jaws possible. They then lost the outer armor and became the sharks, and related organisms, of today. The sharks have no armor and retain a cartilage skeleton (though their teeth are of a bonelike material). They have remained successful to the present day.

In the other direction, the ostracoderms, with limbs and jaws, did not simply get rid of their outer armor, but withdrew some of it under the skin. The armor that had protected the head became a skull to protect the brain and sense organs. Bone spread to the rest of the skeleton, too. In this way, the "osteichthyes" (Greek for "bony fish") developed about 420 million years ago, and still dominate the waters of the earth.

From the bony fish, the amphibians evolved; from the amphibians, the reptiles evolved; from the reptiles, the birds and mammals evolved. All of these have retained the bony internal skeleton. That, of course, includes you. It is your mark as a vertebrate. Nothing that is not vertebrate has bone.

* * *

Bone, like oyster shells, is a calcium compound. How do bones differ from oyster shells, then?

The first person to do a successful chemical analysis of bone was the Swedish mineralogist Johann Gottlieb Gahn (1745–1818). He made use of the then-new method of blowpipe analysis. The blowpipe produced a small, hot flame in which minerals could be heated. The manner of their melting or vaporizing, the colors they formed, the characteristics of their ash could all be interpreted by a skilled practitioner. In 1770, Gahn subjected bone to blowpipe analysis and found it contained calcium phosphate, whose molecule, as you can tell from its name, contains a phosphorus atom.

In the previous chapter, I described how phosphorus had been discovered just a century before Gahn's discovery. It had been obtained from urine, so that suggested it might be a component of the body (or just an impurity that was cast out in the urine as fast as possible). Gahn was the first to point out a specific place in the body where it could be found. It existed in bone.

However, bone exists only in vertebrates. What about nonvertebrate animals? What about plants? Is phosphorus only to be found in one place or might it be a universal component of all life-forms?

In 1804, a Swiss biologist, Nicolas Theodore de Saussure (1767–1845), published a number of analyses of different plants, of the water-soluble minerals they contained, and of the ash obtained when they had been burned. He invariably found phosphates present, which could indicate that phosphorus compounds were a universal constituent of plant life, and possibly of all life.

On the other hand, plants might take up miscellaneous atoms from the soil in which they grew, even a relative few for which they had no use. In that case, since plants did not have the efficient excretory systems of animals, they might simply store the unnecessary atoms in odd corners of their tissues, and they would be there to show up in analysis. Thus, Saussure also discovered small quantities of silicon compounds and aluminum compounds in plant ash and, to this day, we have no clear evidence that either silicon or aluminum are essential components of life.

We might work from the other end and find out what elements contributed to plant growth. It was clear from earliest times that when plants were cultivated, they withdrew vital matter from the soil, and, if this was not restored, the soil gradually became infertile. By hit and miss, various animal products were found to work as "fertilizers"—blood, ground bones, decaying fish, and so on. The most common fertilizer, because it was so handy, was animal (or human) manure. So commonly was it used that, to this day, "fertilizer" is a genteel synonym for manure, which is itself from an old French word meaning "to cultivate" and is a genteel synonym for you-know-what.

The trouble with manure and other animal products is that they are so complex, chemically speaking, that we can't be sure which components do the actual fertilizing because they are essential to plant growth and which are only along for the ride.

In the nineteenth century, however, there was a drive to replace manure. For one thing, manure stinks (as we all know) and makes a travesty of the "fresh air" of the countryside. For another, it carries disease germs, and probably did its bit to initiate and make worse the epidemics that struck the world in the old days.

The German chemist Justus von Liebig (1803–73) was the first to study chemical fertilizers in detail and, by 1855, he had made it quite plain that phosphates were essential to fertilization.

If phosphates are essential to plants and, presumably, to animals, then it must be found elsewhere than in bones. It must be present in the soft tissues, and that means there must be some organic compounds built up of the ordinary elements found in such substances (carbon, hydrogen, oxygen, nitrogen, sulfur), but with phosphorus atoms in addition.

Such a compound was actually found, even before Liebig had worked out his fertilizer system. In 1845, a French chemist, Nicolas Theodor Gobley (1811–76), was studying the fatty matter in egg yolk. He obtained a substance whose molecules he hydrolyzed (that is, broke apart with the addition of water) and obtained fatty acids. This is what is to be expected of any self-respecting fat. He also obtained, however, "glycerophosphoric acid," an organic molecule containing a phosphorus atom. In 1850, he named the original substance "lecithin," from the Greek word for "egg yolk."

Gobley was not able to get the exact chemical analysis, but we now know what it is. The lecithin molecule is made up of 42 carbon atoms, 84 hydrogen atoms, 9 oxygen atoms, 1 nitrogen atom, and 1 phosphorus atom. Only 1 phosphorus atom out of 137 atoms altogether, but that is enough to establish the existence of organic phosphates.

There are other similar compounds that have since been discovered and that are called, as a group, "phosphoglycerides."

Actually, the phosphoglycerides might also be considered skeletal matter. They help make up the cell membranes and the insulating material about nerve cells. In fact, the white matter of the brain (white because of the presence of thick layers of insulating fatty material) surrounding the nerve fibers is particularly rich in phosphoglycerides.

When this was first discovered, it was thought that phosphorus had something to do with mental function, and the slogan arose, "No phosphorus, no thought." In a way, that was right, since if the nerve fibers are not properly insulated, they won't work, and we won't think. That, however, is an indirect connection. We might as well say, since kidneys are essential to human life, "No kidneys, no thought," which is true enough, but which doesn't mean that we think with our kidneys.

It was also discovered that fish is reasonably rich in phosphorus, from which arose the myth that fish is "brain food." This, in popular food mythology, must be second only to the notion (fostered by good old Popeye) that spinach is the gateway to instant superhuman strength. Bertie Wooster, that lovable but dim-witted young man created by P. G. Wodehouse, was always urging his intelligent manservant, Jeeves, to eat a few sardines whenever some particularly urgent problem arose.

Once lecithin was discovered, the dam broke. Other

organic phosphates were found. Phosphate groups were found to be part of proteins in milk, eggs, and meat. Obviously, phosphorus was essential to life itself and not just to the skeletal background.

But what does phosphorus, and all those phosphate groups, *do?* It's not enough just to be there. They must do something.

The first hint in that direction came in 1904, when the English biochemist Arthur Harden (1865–1904) was studying yeast, trying to work out the chemical details of how it fermented sugar into alcohol. It did this as a result of the presence of enzymes, and, in those days, nothing more was known about enzymes than the name (Greek for "in yeast") and the fact that they brought about chemical changes.

Harden placed the ground-up yeast, containing the enzymes, in a bag made of a membrane that was porous enough to allow small molecules, but not big ones, to pass through. After he kept the bag in a vat of water long enough to allow all the small molecules to escape, he found that the material inside the membrane would no longer ferment sugar. That did not mean that the enzyme was a small molecule that had escaped, for the outside water could not ferment sugar either. However, if the material in the vat and in the bag were mixed, the two together could ferment sugar.

In this way, Harden showed that an enzyme consisted of a large molecule (enzyme) working with a small molecule (coenzyme) in cooperation. The small enzyme, Harden found, contained phosphorus.

This meant that phosphorus was involved in the molecular changes that took place in the tissues directly involved. Phosphates were part of coenzymes that worked with many enzymes, and that was not all.

Yeast extract ferments sugar quite rapidly at first, but as time goes on the level of activity drops off. The natural assumption is that the enzyme breaks down with time. In 1905, however, Harden showed that this could not be so. If he added inorganic phosphate to the solution, the enzyme went back to work as hard as ever and the inorganic phosphate disappeared.

What happened to the inorganic phosphate? It had to attach to something. Harden searched and discovered that two phosphate groups had attached to a simple sugar, fructose. The molecule that resulted, "fructose-1, 6-disphosphate," is sometimes called Harden-Young ester, in honor of Harden and his coworker, W. J. Young.

Harden-Young ester is an example of a "metabolic intermediate," a compound formed in the course of metabolism, in places between the starting point (sugar) and the ending point (alcohol). Again, once the first step was taken, others followed and many other phosphorus-containing metabolic intermediates were discovered.

But why should these phosphorus-containing metabolic intermediates be important? The German-American biochemist Fritz Albert Lipmann (1899–1986) glimpsed the answer in 1941. He noticed that most organic phosphates, when they were hydrolyzed and the phosphate group broken off, liberated a certain amount of energy, about the amount one might expect.

On the other hand, when a few phosphate esters were hydrolyzed, they liberated a rather greater amount of energy. Lipmann therefore began to speak of a low-energy phosphate bond and a high-energy phosphate bond.

Food contains a great deal of chemical energy, and when it is broken down it yields more energy all together than the body can easily absorb. There is a danger that most of the energy would be lost. However, as the metabolic chain progresses, every once in a while enough energy is produced to change a low-energy phosphate bond into a high-energy one that contains a convenient amount of energy.

It is as though food consisted of hundred-dollar bills that the body couldn't find change for, but when it was broken down and formed high-energy phosphate bonds, it was as though the hundred-dollar bills were broken up into many five-dollar bills, each of which is easily negotiable.

The most common and ubiquitous of the high-energy phosphate bonds in the body are those belonging to a molecule called "adenosine triphosphate" (ATP), and it is this which is *the* energy-handler of the body. For a few years, ATP was considered the key phosphorus compound of life.

However, as long ago as 1869, a Swiss chemist, Johann Friedrich Mischer (1844–95), isolated, from pus, an organic substance that contained phosphorus. He reported this to his boss, the German biochemist Ernst Felix Immanuel Hoppe-Seyler (1825–95), who was dubious about the worth of the discovery. At that time, lecithin, discovered 24 years earlier, was still the only phosphorus-containing organic substance known, and Hoppe-Seyler wasn't anxious to make a fool of himself by allowing his laboratory to report a second until he was *sure*. (That's responsible science!) After two years he had isolated the substance from other sources, too, and finally came to the conclusion that it was an authentic discovery.

Because cell nuclei seemed to be particularly rich in the substance, it was called "nuclein." Later, as its chemistry was better worked out, it became "nucleic acid."

To sum up, beginning in 1944, nucleic acids came to assume a critical position in the views of biochemists—particularly the variety known as "deoxyribonucleic acid" (DNA), which is now viewed as the key to and the fundamental component of life. It is the blueprint for protein construction, and it is the proteins (especially those that are enzymes) that control the chemistry of the cell and make the difference between thee and me, and between either of us and an oak tree, or an amoeba.

It would probably be oversimplifying the matter, but I am strongly tempted to say, "All life is nucleic acid; the rest is commentary."

(And I can't help but think of Coeurl, the felinoid monster in A. E. van Vogt's great story "Black Destroyer," who lived on a planet from which all the phosphorus had sunk into unavailability—and then he sensed the phosphorus in the bones of the human explorers who had just arrived by spaceship.—And that appeared in 1939, well before the importance of nucleic acids was understood.)

Part II
The Solar System

7

The Moon and We

Sometimes I can foresee a question and I am ready for it.

For instance, some days ago, I found myself involved in a long-distance debate with three other science fiction writers. Two of them were in Sydney, Australia, getting ready to attend the World Science Fiction Convention there. A third was in Auckland, New Zealand, on his way to the same convention. And I was in New York because I don't travel.

The debate was on Reagan's Star Wars. Two of the writers were for it, and two were against it. I was one of those against.

I went to a studio here in New York and at 7 P.M. they began to set up the three-way hookup between New York, Sydney, and Auckland, with people in London helping out. It took a while.

Ordinarily I quickly get impatient and begin snarling at such delays, for each minute makes me more conscious of the fact that I am being kept away from my typewriter. This time, however, I managed to remain calm—even amused—for I was anticipating the first question.

Eventually the hookup was established and, to my delight, the first question was handed to me.

"Mr. Asimov," said the host, "do you think Star Wars will work?"

I answered something like this: "Star Wars is going to involve computers far more complex than any we now have, and a few other devices we haven't developed yet, and a number of processes we haven't worked out at all. When we finally do get it set up—if ever—it will be the most complicated system we have ever worked with and there will be no chance to test it under field conditions until such time as the Soviet Union chooses to launch a mass nuclear attack. At that time, it will have to work, first time, from a cold start, with total accuracy and efficiency, or civilization may be destroyed.

"On the other hand, we've had radio for eighty years, and communications satellites for twenty-five years, and when it came to setting up a hookup with these old, well-established items, it took you thirty-five minutes of fiddling. Honestly, then, do *you* think Star Wars will work, and are you willing to risk the world on it?"

Although I don't like participating in debates, I must admit that I enjoyed that moment.

There are also times, however, when I *don't* anticipate questions and the subject of this essay arose out of one that struck me totally by surprise. It came about this way—

* * *

I write an editorial for each issue of *Isaac Asimov's Science Fiction Magazine* on some subject of science fictional interest.

In the May 1985 issue, I wrote one called "Moonshine" that was inspired by the movies I have seen in which men turn into wolves or take up some form of violent, aberrant behavior on the night of the Full Moon. The implication is that there is something about the light of the Full Moon that exerts a weird influence on the human body. (Of course, the Moon is full only once a month, but in these pictures, the Full Moon appears every other day, on the same principle whereby a six-shooter in the average Western movie pumps out thirty-seven bullets without reloading.)

On a more "scientific" note, however, people are constantly reporting that the rates of murder, suicide, and violent crime in general go up with the Full Moon and, again, the implication is that there is something eerie about the light at that time.

In my editorial, then, I speculated on the possibility that there might truly be some rational explanation for periodic changes in human behavior with the changing of the phases of the Moon.

Surely, no sensible person can believe that moonlight itself can seriously affect human beings. It is, after all, merely reflected sunlight that is somewhat polarized. And even if moonlight had some effect, why should the light of the Full Moon have some effect while that of the first quarter or third quarter, or even the light of the Moon one day after full or one day before full has none. I suppose no one will maintain, seriously, that moonlight on the one night of the Full Moon is so different from other moonlight that it will turn a man into a wolf. For myself, I don't see how such a light would directly influence human behavior in any marked way.

Of course, a person might argue that the effect of moonlight is indirect. On the night of the Full Moon, the night is much more illuminated than at other times and this encourages nocturnal activity and, therefore, night crime. But consider this:

1) The night is pretty well illuminated during the entire week of the Full Moon. It is not much brighter on the actual night of the Full Moon than on the night before or the night after. Why all this fuss about the actual night of the Full Moon, then?

2) The sky is often cloudy, and the night may be very dark even at the time of the Full Moon. Do all the peculiar events that seem to be associated with the Full Moon take place only when the nights are clear? I haven't heard anything like this.

But a person might argue that I haven't studied this "Moon effect" in detail. They may claim that the crime level and other peculiar behavior *does* rise and fall slowly with the level of night illumination and that it *is* more marked when the sky is clear than when it is cloudy. I doubt this, but let's concede it and move on to the next point.

3) People who make a fuss about the phases of the Moon and think that the level of lunar night illumination is of importance are living, at best, in the world of a century ago. We are now living in the era of artificial illumination. Night after night, American cities are so bright that astronomers are going out of their

minds in their search for darkness in which to practice their calling. What ordinary person knows what the phases of the Moon are these days or cares? Moonlight, whether full, part-full, or none, makes no difference to the total light level in any reasonably inhabited locality these days.

It may be, though, that a person might argue that the influence of the Moon is more subtle than that involved in its light. The Moon effect may depend upon something that does not face competition from artificial lighting, and that goes right through any clouds that might exist, and that is at a sharp maximum at the time the Moon is full.

That's a lot to ask, but, as it happens, the Moon *does* exert an effect on the Earth that is quite independent of its light; that does not face competition from anything earthly; and that does indeed pass right through clouds or any other conceivable barrier. It is not a particularly arcane force, however; it is the Moon's gravitational pull.

As a result of its gravitational field, the Moon exerts a tidal effect on the Earth. The tide is low at moonrise and moonset. It is high when the Moon is halfway between setting and rising, whether it is crossing the meridian high in the sky or is as far below the Earth as it can get, on the antimeridian on the other side of the celestial sphere.

What's more, the high tide is higher than usual, or lower than usual, with the changing relative position of the Moon and Sun, since the Sun's gravitational pull also creates tides (though lesser ones than the Moon's). This means that high tides are higher or lower with the changing phases of the Moon, since tidal effects also depend on the relative positions of the Moon and Sun.

At Full Moon and New Moon, the Moon and Sun pull along the same line and the high tides are then at their highest, and the low tides at their lowest. When the Moon is at the first quarter or third quarter, the Moon and Sun are pulling at right angles to each other and the high tides are then at their least high and the low tides at their least low.

In other words there are two tidal cycles. One is a simple up-and-down cycle that repeats itself every half day. Another is a slower rise and fall of the high and low tides themselves and that completes its period in about a month.

The question, then, is whether either of these tidal rhythms can have any effect on human behavior? If there is, the effect certainly isn't one that makes itself consciously felt. Can you tell when high tide or low tide is by the way you feel?

Of course, it may be that the tidal rhythms affect us in ways we can't ordinarily detect. It may affect the hormonal balance in our blood and give us a greater tendency to nightmares, or to irrational rages, or to profound depressions, at certain phases of the Moon.

But how would the tidal rhythm do this? There might be a tendency to speak of unknown forces or influences—but that way lies mumbo jumbo.

To this you might reply, "Nonsense! There was a time, prior to 1801, when ultraviolet light was unknown. Yet it could give you sunburn even in 25,000 B.C."

Suppose a Cro-Magnon man of 25,000 B.C. had said, "I got sunburned through the action of an undetectable component of sunlight." Would that be mumbo jumbo, or would it be a case of remarkable insight?

Well, before you vote for insight, remember that that same Cro-Magnon man could just as easily have said, "I should be made leader of the tribe because an undetectable component of sunlight is filling me with special charisma and divine power that the rest of you don't have."

In other words, if you deal with some unknown, undetected force, you can make it responsible for anything at all, and there would be no way of telling whether some particular statement you make about it is true or false. In fact, since there are many more potentially false statements than potentially true ones (as, for instance, 2 + 2 has one correct answer and an infinite number of wrong answers, even if we confine ourselves to integers), then anything we say about something we know nothing about is almost certain to be wrong.

To hide behind the unknown, then, is virtually certain to lead us astray and we cannot do that and still be playing the game of science.

But people might say, "We're not talking about an unknown force. We're talking about tidal effect. The tides make themselves felt in the ocean, which is a vast solution of salt water. Human tissue is made up mostly of salt water. Naturally, the tides affect us the way they affect the ocean, so that when we're talking about the Full Moon, we're talking about high tide in the human body."

The tides are just as high at New Moon, but somehow it's always Full Moon people talk about. Still, forget that for a moment, and let's make another point.

The tidal effect is felt by the entire Earth. There are tides in the atmosphere and in the solid outer layers of the Earth as well as in the ocean. It just happens that the ocean tides are more noticeable to casual observation. We therefore can't blame anything on the watery nature of human tissue.

Whereupon you might say, "That doesn't matter. If the tides affect the entire human body, that's all the better."

Let's make another and more important point, then.

The tidal effect is produced by the variations in the gravitational pull from place to place. It varies in intensity with the square of the distance from the body exerting the pull. The side of the Earth nearest the Moon feels the pull more strongly than the side of the Earth away from the Moon. The far side, after all, is 12,756 kilometers farther from the Moon than the near side is. The Earth is stretched by this difference in pull and that produces the small bulges on either side, toward and away from the Moon, and these are the tides.

If we were dealing with a body smaller than the Earth, the difference in the distance from the Moon of the near side and the far side would be smaller, and the tidal effect would be smaller, too, by the square of the extent of the difference in size.

A human being standing upright under the Moon, when it is high in the sky, would have his feet about 1.8 meters farther from the Moon than his head is. That means the Earth is just over seven million times as thick as a human being. Square that and what we are saying is that the Moon's tidal effect on a human being is about 1/50,000,000,000,000 (one fifty-trillionth) that on the Earth.

Can such an infra-tiny tidal effect produce any noticeable difference in behavior in a human being?

Well, searching for *something,* here is what I said in my editorial:

> . . .it is certain they [the tide cycles] affect creatures who spend their lives at or near the seashore. The ebb and flow of the tide must be intimately involved with the rhythm of their lives. Thus, the time of highest tide may be the appropriate occasion to lay eggs, for instance. The behavior of such creatures therefore seems to be related to the phase of the Moon. That is not mysterious if you consider the Moon/tide/behavior connection. If, however, you leave out the intermediate step and consider only a Moon/behavior connection, you change a rational view into a semi-mystical one.
>
> But what connection can there be between worms and fish living at the edge of the sea, and human beings?
>
> Surely there is an evolutionary connection. We may consider ourselves far removed from tidal creatures *now* but we are descended from organisms that, 400 million years ago, were probably living at the sea-land interface and were intimately affected by tidal rhythms.
>
> Yes, but that was 400 million years ago. Can we argue that the tidal rhythms of those days would affect us now? It doesn't seem likely, but it is a conceivable possibility.
>
> After all . . . we still have a few bones at the bottom end of our spine that represent all that is

> left of a tail that our ancestors haven't had for at least 20 million years. We have an appendix that is the remnant of an organ that hasn't been used for even longer. . . .
>
> Why should there not also be vestigial remnants of ancestral biochemical or psychological properties? In particular, why should we not retain some aspects of the old tidal rhythms. . . ?

In this way, I built up an argument that tidal rhythms might affect us as vestigial remnants of behavior dating back to ancestors for whom they were life and death. That, however, only supplies a rational skeleton upon which to hang this business of the "Moon effect." We have to make hard and fast observations about, for instance, the rise and fall of hormonal concentrations with the tides and demonstrate just how this may affect behavior. Without that, all we have is anecdotal evidence, which is notably untrustworthy.

In my editorial, I thought I had taken up the matter in careful and objective detail (as I have here—even more carefully and objectively), but then I got letters of a kind I had never expected asking a question that caught me completely by surprise.

Why, those letters asked, did I neglect to mention the obvious connection between the Moon and menstruation?

What's more, the tone of those letters (all from women, by the way) was personally horrifying. They definitely seemed to think that I had a sexist motive in not discussing the matter; that I simply thought that since menstruation was a phenomenon of females exclusively, I dismissed it as not worth mentioning. More than one letter accused me of "forgetting" 51 percent of the human race.

Why didn't I mention menstruation, then? Simply because it never occurred to me that anyone who thought about it at all would connect it with the Moon.

To be sure, the menstrual cycle in human females does indeed seem to be about the length of the cycle of the phases of the Moon. The correspondence is noticeable enough so that the word "menstruation" is from the Latin "mensis," meaning "month."—But of what value is that? We call native Americans "Indians" because Columbus thought that he had reached the Indies, but the fact that that is what we call them is no evidence that the United States is part of India.

In this connection, consider that of all animals, only the primates menstruate. The menstrual period varies considerably among the various species of primates, so that human beings are one of, at best, very few species that have a menstrual period that is about a month long. If we want to blame that period on the Moon, we're going to have to explain why the Moon's influence is so finely focused. Why does the Moon pick on human beings to the almost total exclusion of all other species?

Then, too, when particular species are affected by some cycle, all the individuals react in about the same way. When one tree of a particular species in a particular region begins to put out leaves in the spring, all the others do so at about the same time. When one swallow returns to Capistrano, so do all the others.

We might expect, then, that being affected by the phases of the Moon, either through tidal rhythms, or in some other way, all women would experience the onset of menstruation at some particular lunar phase. This, however, is not so. There isn't a day in the year when a little under 4 percent of women of the proper age and condition don't experience the onset of menstruation. The phase of the Moon doesn't matter.

To be sure, I've heard that if a group of women are kept in close quarters, their periods tend to begin to match and fall in step. Presumably, they affect one another. Perhaps there is a subtle menstrual odor that tends to stimulate onset. But even so, if it does happen, I've never heard that the matching onsets always come at some particular phase of the Moon. They can line up at *any* phase, apparently.

In that case, we might argue, it isn't the details of the period that have anything to do with the Moon; it is merely the *length* of the period that has something to do with the Moon.

To be sure, I'm a male and have no personal experience of the menstrual period, but I'm reasonably observant and I'm well aware that women are always being surprised by a period beginning a day, or two, or three, ahead of schedule; and of being exalted (or scared, depending on circumstances) because it has been delayed a day, or two, or three.

In short, I'm afraid that the length of the menstrual period is quite irregular in a Universe in which the cycle of the Moon's phases is very regular.

But I can hear someone say, "Never mind the irregularities. The *average* length of the menstrual period is 28 days, and that's the length of the cycle of the Moon's phases, and therefore of the tidal rhythms."

Well, I'm sorry, but that is not the length of the cycle of the Moon's phases, and I will explain why.

The Moon revolves about the Earth (relative to the stars) in 27.3216614 days, or 27 days 7 hours 43 minutes 11.5 seconds. We can call it 27⅓ days without being too far off at all. This is called the "sidereal month," from a Latin word for "constellation" or "star."

The sidereal month is, however, only of interest to astronomers, for it has nothing to do with the phases of the Moon, and it is by the cycle of the phases that ancient peoples defined the month.

The phases depend on the relative positions of the Moon and the Sun. It is the period from New Moon to New Moon, when the Sun and Moon are as close as possible in the sky, so that the Sun and Moon both cross the meridian at noon; or from Full Moon to Full Moon, when Sun and Moon are in directly opposite positions in the sky, so that the Sun crosses the meridian at noon while the Moon crosses it at midnight.

To find this period we have to imagine the Moon starting with the Sun and moving around the sky till it is back at the Sun (New Moon to New Moon). But since the Moon circles the Earth in 27⅓ days, isn't it back at the Sun in 27⅓ days? No, it isn't, because the Sun isn't standing still.

The Earth revolves about the Sun in 365.2422 days and that causes the Sun to seem to move west to east across the sky (relative to the stars). If the Moon starts with the Sun and moves west to east, returning to the same spot (relative to the stars) 27⅓ days later, the Sun has moved somewhat eastward in that period and the Moon must spend extra time catching up to the Sun in order to be at New Moon again. This extra time turns out to be about 2⅕ days, so that the average period from New Moon to New Moon is 29.5305882 days, or 29 days 12 hours 44 minutes 2.8 seconds. We can call it 29½ days and not be very far wrong.

The 29½-day period is called the "synodic month," from a Greek word for a religious gathering, because it was usually left to the priests to decide when the New Moon took place so that the new month could begin at the proper time and with the proper ritual.

The menstrual period is 28 days long, however, while the period of the phases of the Moon is 29½ days long. Isn't that close enough? They're almost equal.

No, it isn't close enough. If the Moon's phases and the tidal rhythm has some connection with the menstrual rhythm, then the two rhythms ought to match—but they don't.

Suppose that someone with a perfectly regular menstrual period and with a perfectly average length of period experiences the onset of that period on a certain day on which the Full Moon shines down on the Earth. If there is any meaning to the myth of the Moon's connection with the menstrual period, then the next onset should be at the time of the next Full Moon and the one after that at the time of the Full Moon after that, and so on and so on indefinitely.

But that doesn't happen! Someone with a perfectly regular period of perfectly average length will experience the next onset one and a half days before the Full Moon, and the one after that three days before the

Full Moon, and the one after that four and a half days before the Full Moon.

Gradually, a woman will experience onset at slightly different phases of the Moon as she works her way through the cycle of phases in a little less than twenty menstrual cycles, and even so the twentieth cycle won't start exactly on the day of the Full Moon either.

Fifty-nine successive perfectly regular menstrual periods will take 1,652 days (or a little over four and a half years). Fifty-six synodic months will also take 1,652 days. Those are the smallest numbers of the two cycles that match each other. That means that, counting from a first onset that takes place on the night of the Full Moon, it will not be until four and a half years later, on the fifty-ninth onset, that the Full Moon and menstrual onset will occur together again.

On the whole, then, no matter how you slice it, the Moon and menstruation have no significant connection in any way.

But then, how do I explain the fact that the menstrual period is so close to the length of the synodic month, if the Moon is not involved?

Well, there *is* an explanation for that, but it is a very undramatic one and many people may not be able to bring themselves to accept it. It's called "coincidence."

I wonder, now, whether this view will offend some women who read this essay. Is there any reason why they should *want* there to be a connection between their bodily processes and the Moon?

Perhaps there is. Perhaps it gives them a feeling of importance to imagine a connection with the Moon that men don't have.

However, that's a nonexistent connection, and it's my own opinion that women are quite wonderful enough in their own right to require no support from a superstitious illusion.

8

The Unmentionable Planet

Back in the 1950s, I wrote a series of six adventure books for young readers, featuring a young man named Lucky Starr. Each book was set in a different portion of the Solar System. In order, the settings were Mars, the asteroids, Venus, Mercury, Jupiter, and Saturn. A seventh book (never written) was to be set on Pluto. I don't think I ever considered using Uranus as a setting, however.

Uranus seems to be the least regarded of the planets. Every other planet has something remarkable about it that makes it a logical setting for science fiction stories. Mercury is the closest to the Sun, Venus is the closest to the Earth, Mars is the best known, Jupiter is the largest, Saturn has its rings, Neptune is the farthest giant; Pluto is the farthest planetary object of any size that can be observed throughout its orbit.

But Uranus? What about Uranus? Is it neglected only because it has nothing remarkable about it? Surely not! I think that, in part, it is because it has the handicap of an unfortunate name; one that, in the English language, at least, is all but unmentionable.

I learned this the hard way. In my young and carefree days, I knew that Uranus (more properly, Ouranos) was the Greek god of the sky. I knew that the muse of astronomy was, therefore, Urania (yoo-RAY-nee-uh). I also knew that there was an element that had, upon its discovery, been named "uranium" for the then newly discovered planet Uranus, and that this was pronounced "yoo-RAY-nee-um."

It seemed to me obvious, therefore, that the name of the planet was pronounced "yoo-RAY-nus" and it was in this way that I pronounced it. So sure was I, that I never bothered to check the dictionary. Not once, moreover, was I wise enough to notice that the name of the planet, so pronounced, was homonymous with "your anus."

The time came, however, when someone pronounced it with the accent on the first syllable. I at once corrected him, with my usual insufferable air of superiority, and, in the argument that followed, we referred to the dictionary and, to my horror, I lost. The victor, dissatisfied with mere victory, squashed me flat by informing me of the distasteful nature of my pronunciation.

As it happens, though, Uranus pronounced "YOO-rih-nus" isn't much better, for it is then homonymous with "urinous," meaning "to have the appearance, properties, or odor of urine."

The result is that the two possible alternate pronunciations of the planet's name are each unpalatable in English and so people avoid mentioning the planet. I know the solution, of course. Either the Greek version of the name, Ouranos (pronounced "OO-rih-nus"), can be adopted, or else the word can be pronounced with a short "a" (Yoo-RAN-us). Since these suggestions are sensible ones, they will never be adopted.

In early 1986, however, Uranus was much in the news, and people had to say its name. So now, with that in mind, I will deal with the planet. I had discussed it in earlier essays in this series but everything prior to January 1986 is ancient history now as far as Uranus is concerned.

In 1977, two probes, *Voyager 1* and *Voyager 2,* were launched, and sent toward Jupiter and Saturn in order to study those giant planets. They passed Jupiter in 1979 and Saturn in 1980, and performed well. After that, *Voyager 1* moved out of the planetary plane and wandered off indefinitely through the spatial abyss.

Voyager 2, however, was redirected in its flight so that it might pass close to the still farther planets, Uranus and Neptune. What's more, the devices it carried were punched up in a series of clever moves so that when it should finally reach Uranus it would be better equipped to study the planet than it would have been in its freshly launched state in 1977.

Uranus is distinctly smaller than Jupiter or Saturn. Uranus's diameter of 51,000 kilometers (32,000 miles) is only about three-sevenths that of Saturn and about one-third that of Jupiter. It is, nevertheless, six and a half times that of Earth, so it is still a "gas giant." Looking at it another way, Uranus has two-thirteenths the mass of Saturn and one twenty-secondth the mass of Jupiter, but it has fourteen and a half times the mass of Earth.

Most planets have axes of rotation that are more or less at right angles to their orbital planes about the Sun. In other words, if we look at the planet in the sky its axis of rotation is close to being up and down. There is usually a small tilt. Venus and Jupiter have axes that are 3 degrees from the upright; Earth has a tilt of 23.5 degrees; Mars, one of 24 degrees; Saturn, nearly 27 degrees; Neptune, nearly 29 degrees. Mercury's axial inclination is uncertain but it is something less than 28 degrees.

Presumably, if the planetary system had formed from a vast cloud of dust and gas, swirling about in eddies and subeddies, all the planetary axes should have been exactly perpendicular to the plane of the orbit about the Sun.

The planets were formed, however, by the coming together of subplanetary objects. If those objects came together from all directions equally, then their axes would remain perpendicular. However, it is likely that the directions of the later and larger slams were not equally balanced, so that their axes would be somewhat knocked out of true by a random amount.

Well, Uranus must have received one or more terrific knocks while it was forming and, as luck would have it, from more or less the same direction, for its axis is tilted by a colossal 98 degrees—a little more than a right angle.

This means that Uranus is rotating about its axis on its side and in viewing the planet in the sky, we see the axis extend from left to right rather than up and down.

Uranus revolves about the Sun in 84 years and, because of the tipping of its axis, the northern hemisphere sees the Sun spiraling up to zenith and then down to the horizon during half the revolution, while the southern hemisphere sees that happen during the other half.

If one were at Uranus's north pole (or south pole), one would see the Sun rise at some point on the horizon, then slowly ascend the vault of the sky until, after some 21 years (!) it would be nearly overhead. It would then descend for another 21 years, finally setting at the opposite point on the horizon after having remained in the sky for 42 years. It will then be 42 years more before it rises again.

A human being born at one of Uranus's poles would be a middle-aged man at sunset and a very old man before it was time for a second sunrise.

At the present moment, in the skies of Uranus, the Sun is nearly at zenith over the planet's south pole. In other words, the south pole is pointing nearly directly toward the Earth and Sun. (It has to point at both, for, from Uranus, the Earth is never seen farther than 3 degrees from the Sun.)

As 1985 drew to its close, *Voyager 2* was approaching Uranus and getting ready to take its photographs and make its measurements. It had traveled about 10½ billion kilometers (6½ billion miles) to do this. Uranus is only 2¾ billion kilometers—1⅔ billion miles—from us as the crow flies—however, *Voyager 2* was not flying in a straight line, but in wide arcs in response to the gravitational pull of the Sun, Jupiter, and Saturn, and to the original motion of the Earth at the time of launch.)

Having come all that way, *Voyager 2* found itself in dim surroundings. The intensity of light from Uranus's very distant Sun is ¼ that at Saturn, 1/13 that at

Jupiter, and 1/368 that at Earth. This means that photographs at Uranus had to be of longer exposure than those taken at Jupiter and Saturn. At Saturn, exposures of 15 seconds were sufficient, but at Uranus exposures of nearly 100 seconds were needed. This means there was time for fewer photographs and that there was a greater chance of fuzziness.

In appearance, Uranus was bluish and nearly featureless. This was not entirely unexpected. The farther from the Sun, the less heat is delivered to the planet and the smaller is the difference in temperature between the various parts of the planetary surface. It is the temperature difference that drives the atmospheric circulation and produces visible clouds and storms.

Hence, Jupiter's atmosphere is banded and tortured, Saturn's is less so, and Uranus's is virtually quiet.

Furthermore, various gases are frozen out of the atmosphere as one recedes from the Sun. Jupiter's atmosphere is comparatively rich in ammonia, plus other gases with comparatively high boiling points, and it is these which help form the clouds and colored formations. On Saturn, the ammonia is lower in the atmosphere (where the temperature rises to the point of keeping it gaseous) and on Uranus, lower still.

This means that on Uranus, methane, which has a particularly low boiling point, is the predominant impurity in the upper atmosphere. Methane absorbs red light and gives the atmosphere a blue appearance. Furthermore, methane tends to undergo chemical reactions in the presence of even the feeble sunlight that bathes that distant planet. This produces a hydrocarbon smog that keeps us from seeing far into the planetary atmosphere. (It is this same sort of smog that exists in the methane-rich atmosphere of Saturn's moon, Titan.)

The chemical reactions undergone by methane may make the gas's presence evident through a change in color in the atmosphere. If so, that change would be most noticeable at the south pole now, where the feeble Sun at zenith delivers slightly more heat than elsewhere; and, indeed, a trifling increase in redness has been reported at the south pole.

Of course, methane is present in the Uranian atmosphere only as a minor component. The chief components (as in the case of Jupiter, Saturn, and, for that matter, the Sun) are hydrogen and helium, with hydrogen predominant.

Recent infrared studies from Earth's surface seemed to show that the Uranian atmosphere might be as much as 40 percent helium. This sent a shudder of dismay through the astronomic fraternity, since the figure is far too high. The helium content of the Universe, generally, is 25 percent by mass, with hydrogen making up the remaining 75 percent (and everything else being less than 1 percent).

The Sun, Jupiter, and Saturn have helium contents of 25 percent or less and it would be a chore indeed to account for the accumulation of helium on Uranus.

One might argue that Uranus being farther from the Sun would have less material out of which to form. It would therefore develop more slowly, and be smaller than Saturn (which is in turn smaller than Jupiter). Because Uranus would be smaller than the two inner gas giants at every stage of its formation, it would have a less intense gravitational field and would gather less hydrogen than Jupiter and Saturn. It might manage to gather the more massive atoms of helium efficiently, however, and in this way it would accumulate not more helium, but at least a higher percentage of it.

The trouble with that notion is that Uranus is colder than either Jupiter or Saturn and, at its lower temperature, it should be able to hold on to hydrogen easily despite its smaller size.

To the infinite relief of astronomers, however, *Voyager 2* eliminated the problem. Its observations showed the helium content of Uranus's atmosphere to be about 12 to 15 percent—exactly where it ought to be.

A total of four clouds were detected fairly deep in the atmosphere, and they were studied carefully in order to determine the rotation period of the planet.

There was a general feeling among astronomers that the smaller the planet, the longer the rotational period. Thus Jupiter, the largest planet, rotates in 9.84 hours; Saturn, the next largest, in 10.23 hours; and Earth, in 24 hours. Uranus, lying between Saturn and Earth in size, ought to have an intermediate rotation period, too.

The usually accepted period of rotation for Uranus, until recently, was 10.8 hours. In 1977, however, a new measurement yielded a rotational period of perhaps as much as 25 hours.

The trouble was, of course, that there was no clear marking on Uranus that could be seen from Earth and followed in its travels. *Voyager 2,* however, presented results that showed that Uranus rotates on its axis in 17.24 hours, which is certainly an acceptable figure.

There are some puzzles about the atmosphere, of course. The temperature at the visible surface of the Uranian atmosphere is about the same everywhere.

The weak solar radiation doesn't seem to make much difference. There is, however, a region at about 30 degrees north latitude and south latitude where the temperature seems to drop a small bit. No acceptable reason has yet been presented for this.

Then, too, winds have been detected in the atmosphere which travel at about a hundred miles an hour in the direction of the planetary rotation. This is puzzling, because what we know of atmospheric motions leads us to suppose that the wind should blow in the direction opposite to that of the planetary rotation. However, Uranus (like Jupiter and Saturn) seems to radiate more energy than it receives from the Sun, so there must be some internal source of heat, some physical or chemical change, that may possibly account for the anomalous wind movement.

As *Voyager 2* approached Uranus, it seemed at first that Uranus had no magnetic field. This was a major shock, for one expected a field if a planet had a rapid rotation and an electrically conductive interior. Since Jupiter and Saturn have magnetic fields, it seemed certain Uranus would have one. If Uranus lacked one, that would call for some tall explaining.

Fortunately, astronomers were rescued. *Voyager 2* was approaching from the sun side and the observation of the magnetic field was blocked by electrons in Uranus's ionosphere. When *Voyager 2* reached a point 470,000 kilometers (290,000 miles) from Uranus's center, they passed into the planetary magnetosphere. The magnetic field existed; it was 50 times as strong as Earth's and it stretched way out on the nightside. All was as it should be.

Well, not all. The magnetic axis is usually tipped with respect to the rotational axis, and the magnetic axis does not necessarily pass through the gravitational center of the planet. (No satisfactory explanation for this has yet been reached.)

The situation in Uranus's case is extreme, however. The magnetic axis is tipped no less than 60 degrees to the rotational axis, and the center of the magnetic axis is 8,000 kilometers (5,000 miles) from the center of the planet. Why this extraordinary displacement should be, we don't know, but perhaps it has something to do with the just as extraordinary tipping of the rotational axis.

Voyager 2 passed between the planet's rings and its innermost satellite (as observed from Earth), Miranda. At 5 P.M. on January 24, 1986, it made its closest approach to Miranda, reaching a point 28,000 kilometers (17,400 miles) from Miranda's surface. Not quite an hour later it made its closest approach to Uranus, when it reached a point 81,500 kilometers (51,000 miles) from Uranus's cloud layer. It made these close passes within seconds of the scheduled time and only 16 kilometers (10 miles) from the scheduled path. That was pinpointing with a vengeance.

Nine thin rings had been detected around Uranus in 1977 by studies from the Earth's surface. This was done by studying Uranus when it passed very close to a star, and noting the way in which the star blinked on and off as the rings passed before it.

Voyager 2 showed the nine rings were really there, and detected a tenth ring between the eighth and ninth, as one counts outward from Uranus. The new ring is very thin and faint and could not possibly have been spotted from Earth.

As was observed even from Earthside observations, Uranus's rings are composed of dark particles. This is, perhaps, no mystery. The smaller bodies in the outer Solar System tend to be icy, where the ice (usually water ice, but perhaps with ammonia and methane as minor components) is mixed with stony materials of various sizes.

Two things can happen to such icy objects that would serve to darken them. They can slowly lose ice through evaporation while not losing the stony materials. With time, through billions of years, small objects become less icy and tend to become covered with a crust of stony material that is darker than the ice and that prevents further ice from evaporating. Secondly, the methane in the ice may slowly polymerize into black, tarry substances that will further darken the surface.

The possible development of such a crust on comets will be mentioned by me in Chapter 10, which was written well before a spaceprobe called Giotto skimmed past Comet Halley. When Giotto made its pass it showed that Comet Halley was very black in color. (It was still sending out jets of evaporating ice, however, for it was heated far more intensely here in the inner Solar System than objects in the neighborhood of Uranus would be.)

The problem, then, is not why Uranus's rings are so dark, but why Saturn's rings are so white. Apparently the small bodies near Saturn (with the exception of the satellite Iapetus, which seems to be darkened over one hemisphere) are icier than those near either Jupiter or Uranus, and this will someday have to be explained.

Again, it turns out that whereas Saturn's rings are composed of particles of all sizes, from fine dust to what are almost mountains, Uranus's rings are composed of objects that are comparatively uniformly boulder-size. Uranus's rings are virtually dust-free. This is again an unexplained difference between Saturn and Uranus and it is my guess that it will be Saturn that will turn out to be atypical.

Uranus has a satellite system that is peculiar in several ways. Five satellites were discovered from Earth, and of these five, none are giant satellites with diameters of 3,000 kilometers (1,850 miles) or more. Uranus is the only gas giant without a giant satellite. Neptune has Triton, Saturn has Titan, and Jupiter has Io, Europa, Ganymede, and Callisto. Even the Earth has the Moon. Why Uranus should lack a giant satellite we don't know. Can it have something to do with Uranus's extraordinary axial tip?

The five satellites are as tipped as Uranus, by the way, and revolve in Uranus's equatorial plane. That means that while all other planetary satellites move more or less left, right, left, right as we watch them in the sky, the Uranian satellites move up, down, up, down.

This may mean that the satellites were formed after Uranus's axis was tipped. If Uranus had had a relatively untipped axis with the satellites in place in its then equatorial plane, the tipping of Uranus would have left the satellites moving in highly inclined orbits. The tipping must have taken place very early in the history of the Solar System and the satellites came into being afterward.

The satellites are darker than expected. They could not be made out as anything but dots of light from Earth, so astronomers judged their size from their brightness, assuming a moderately high ability to reflect light, since they were thought to be icy. Since they proved darker than had been thought, they must reflect less light and must be larger in order to be as bright as they appear. Here is a list of the diameters of the five satellites, as they were thought to be before *Voyager 2*, and as they are known to be now:

Satellite	Diameter	
	kilometers (miles)	
	Before *Voyager 2*	After *Voyager 2*
Miranda	240 (150)	480 (300)
Ariel	700 (435)	1,170 (725)
Umbriel	500 (310)	1,190 (740)
Titania	1,000 (620)	1,590 (990)
Oberon	900 (560)	1,550 (965)

Notice that Miranda's diameter has been increased 2.0 times, Ariel's 1.7 times, Umbriel's 2.4 times, Titania's 1.6 times, and Oberon's 1.7 times. Naturally, the satellites were discovered in order of decreasing size. Miranda, the smallest of the five (and the innermost) was not discovered till 1948.

On December 31, 1985, however, the approaching *Voyager 2* discovered a sixth satellite, one that was closer to Uranus than Miranda is. Miranda is 130,000 kilometers (80,400 miles) from Uranus's center, whereas the new satellite is only 85,000 kilometers (53,400 miles) away. The new satellite is also only 160 kilometers (100 miles) in diameter. Its provisional name is 1985U1.

In January 1986, no fewer than nine more satellites were discovered, all closer to Uranus than 1985U1. The first three discovered, 1986U1, 1986U2, and 1986U3, were about 80 kilometers (50 miles) across, the rest between 20 and 50 kilometers (10 and 30 miles) across. The innermost now known is 1986U7, which is only 50,000 kilometers (30,500 miles) from Uranus and is within the ring system.

These small satellites bring with them a couple of problems. The study of Jupiter and Saturn by probe introduced the notion of "shepherd satellites," small satellites that circle on the outside and inside of a particular ring and, by gravitational influence, keep the rings from spreading out and dissipating. Well, most of Uranus's rings do not seem to have such satellites. How, then, do they stay in existence?

Then, too, we find that Jupiter, Saturn, and Uranus all have small satellites circling in or just outside the ring system. Neptune probably has them, too. Mercury and Venus have no satellites at all, and Earth has a large, distant Moon, but no small satellites close in. Is it the absence of these small close-in satellites that keeps these worlds from retaining rings? Mars does have two small satellites close in, but no ring. Were the Martian satellites captured after a ring had dissipated? We're going to have to learn a lot more about ring formation, if we can.

The five comparatively large satellites of Uranus were examined. Oberon has a cratered surface, with bright rays spreading outward from them. That's common enough. The crater floors are dark, however, and that's less common.

Titania has not only craters, but rift valleys. Let's skip Umbriel for the moment and pass on to Ariel, which has even larger rifts and canyons. Apparently, the closer to Uranus, the more tortured the satellite surface.

Miranda, which was seen at closest quarters, was a huge surprise. Its supertortured surface had a little bit of everything. It had canyons like Mars, grooves like Ganymede, sunken terrains like Mercury. In addition, it had a series of dark lines, like a stack of pancakes edge-on, a set of grooves marked out like a racetrack, and a bright, *V*-shaped chevron.

It seemed utterly confusing to have so small a body with such a variety of surface features. It is far too small to be geologically alive. The current speculation is that it suffered near-death. It was struck by some large body, perhaps, and actually shattered. (Saturn's innermost sizable satellite, Mimas, has a crater so large that the impact must have nearly shattered it.)

The shattered Miranda must have come together again under its own gravity, but not in an orderly fashion. It came together every which way and now presents its surface in the chaotic fashion it does.

* * *

It seems to me that the real mystery, however, is Umbriel. It is the darkest of the satellites. It is also apparently featureless, except for one bright ring, like a doughnut shining near the edge of the lighted hemisphere.

Why is Umbriel darker than the rest? Why is it featureless? What produces the white doughnut? Unfortunately, it will probably be many years before we get another (and, perhaps, better) look at Umbriel, and till then we can only stare at the pictures we have and wonder.

But to me, the most interesting thing about Umbriel is a peculiar and assuredly meaningless coincidence.

In 1787, the German-English astronomer William Herschel (1738–1822), who had discovered Uranus six years before, discovered its two brightest satellites. Instead of naming them after Graeco-Roman mythological characters, he called them "Oberon" and "Titania," the King and Queen of fairies in William Shakespeare's *A Midsummer Night's Dream.*

When the English astronomer William Lassell (1799–1880) discovered the third- and fourth-brightest satellites of Uranus in 1851, he named the brighter of the two "Ariel" after the happy, lighthearted spirit in Shakespeare's *The Tempest.* (It was naturally assumed that Ariel, the brighter, was also larger than the other, but we now know that the other is really larger, but darker, so that it reflects less light.)

The other, dimmer, satellite Lassell named "Umbriel," after a spirit in *The Rape of the Lock,* a mock epic written by the English poet Alexander Pope (1688–

1744). Umbriel was a moody spirit, full of sighs and melancholy. He was named Umbriel from the Latin word for "shadow." (Thus, an "umbrella" casts a "little shadow," within which we remain dry.)

When the fifth satellite of Uranus was discovered by the Dutch-American astronomer Gerard Peter Kuiper (1905–73), he went back to *The Tempest* and named the newly discovered object Miranda, after the play's charming heroine.

But isn't it strange that the dark, shadowy satellite, Umbriel, should be named for a moody spirit that sat glumly in the shadows? Is there some deep significance here?

No! Not at all. It's just coincidence.

9

The Incredible Shrinking Planet

A few weeks ago, I received a phone call from a young woman who said she was putting together some sort of article for some sort of magazine. (I don't suppose she was actually a writer, since all she was doing was calling various celebrities in order to ask them a question. She was then going to put all the answers together and have the result appear in print. It doesn't take much writing ability to do that.)

I said, cautiously, "What is the question?"

"Well," she said, vivaciously, "What is your favorite bar and why is it your favorite? Is it because of the quality of its drinks, its ambience, its inaccessibility, the people you find there, or what?"

"My favorite bar?" I said, astonished. "You mean a bar where people go to drink?"

"Yes. Of all the bars you've visited—"

"But I don't visit any, miss. I do not drink. I never

have. I don't suppose I ever enter any bar except to pass through it on my way to a dining area."

There was a pause, and then the caller said, "Aren't you Isaac Asimov, the writer?"

"Yes, I am."

"And aren't you the one who's written about three hundred fifty books?"

"Yes, I have; but I've written every single one of them while stone cold sober."

"You have? But I thought all writers drank." (I think she was being polite at this point. I think that what she really meant to say was that she thought all writers were alcoholics.)

I said, rather stiffly, I suppose, "I can't speak for anyone else, but I don't drink."

"Well, that's certainly strange," muttered the caller, and hung up the phone.

Frankly, I think it did my caller a great deal of good to experience something strange. We should each of us be subject to such a stirring up for the sake of our mental health, and scientists, of course, are fortunate enough to experience it all the time. Take the case, for instance, of the planet Pluto—

Throughout the first third of this century, the search was on for a "Planet X," one with an orbit beyond that of Neptune. Those astronomers who searched for it expected to find a gas giant—that is, a planet that was larger than the Earth, but was low in density because it was made up largely of hydrogen, helium, neon, and the hydrogen-containing "ices," water, ammonia, and methane. After all, the four outermost planets, Jupiter, Saturn, Uranus, and Neptune, were all gas giants, so why shouldn't the planet beyond Neptune be one?

Naturally, astronomers expected that Planet X would be smaller than the known gas giants because it was farther out from the Sun. The farther out, the thinner and more tenuous the preplanetary nebula would have been, and the smaller the planet that would have formed. Even so, Planet X was expected to be substantially larger than Earth.

After all, the mass of Jupiter, the largest gas giant and the nearest to the Sun, is 318 times that of Earth. Saturn, the next farthest out, has a mass 95 times that of Earth. Beyond those two giants are Uranus and Neptune, which have masses of 15 and 17 times that of Earth, respectively.

The American astronomer Percival Lowell (1855–1916), the most assiduous of the searchers, therefore guessed that Planet X would continue the downward trend and might have a mass only 6.6 times that of Earth. Still, no one would have been surprised if it had proved to be as much as 10 times the mass of the Earth.

Moreover, it was not necessary to reason the mass entirely by analogy. There was a stronger argument. The reason why Planet X was thought to exist was because of the slight anomalies in the orbit of Uranus. That meant that astronomers were looking for a planet massive enough to affect Uranus's orbit measurably, even though such a planet might well be two or three billion kilometers beyond Uranus. Having Planet X ten times the mass of the Earth would make it none too large for the job.

Planet X was finally discovered in 1930 by the

American astronomer Clyde Tombaugh, who called it Pluto, partly because the first two letters were the initials of Percival Lowell. It was discovered fairly near the place where it should have been if it were indeed affecting Uranus's orbit and that, too, was a point in favor of the suggestion that it ought to be a gas giant.

The very instant of discovery, however, produced a nasty shock, the first of a number that Pluto would provide over the next half century.

You see, Neptune is an eighth-magnitude object. That makes it too dim to see with the unaided eye, but that is to be expected, considering that it is about 4,500 million kilometers (2,800 million miles) from the Sun and that its reflection of the dim sunshine it receives must then travel that distance again to reach us.

Pluto, allowing for its greater distance and its presumably smaller size, should naturally have been substantially dimmer than Neptune. Astronomers expected Pluto to have a magnitude of perhaps ten.

But that was not so. Pluto was of the fourteenth magnitude. It was almost forty times dimmer than it was expected to be.

There were three possible reasons for this: 1) Pluto was considerably more distant than expected; 2) Pluto was made of considerably darker materials than expected; and 3) Pluto was considerably smaller than expected. Or, of course, there could be any combination of these three possibilities.

The distance was fairly easy to determine. From Pluto's shift in apparent position from day to day, one could get a rough notion pretty quickly as to the time it would take to move about the Sun. From this orbital period, it was at once possible to calculate its average distance from the Sun.

As it has turned out, it takes Pluto 247.7 years to go around its orbit once, and its average distance from the Sun is about 5,900 million kilometers (3,670 million miles). It is, on the average, then, about one and a third times as far from the Sun as Neptune is.

This makes Pluto the most distant of the known planets, to be sure, but it does not place it so far off that one might account for its dimness as a result of distance alone. It follows that Pluto must be made of darker materials than the four gas giants are, or that it is considerably smaller than they are, or both.

Whether one or the other or both, Pluto is *not* a gas giant. For one thing, a gas giant (or any planet with an atmosphere dense enough to produce heavy clouds) reflects about half the sunlight that falls upon it. Its "albedo," in other words, is in the neighborhood of 0.5. The same is true of a planet, even without an atmosphere, if it has an icy surface (one consisting of frozen water, ammonia, methane, or any combination of these). A planet without an atmosphere and consisting of bare rock would have an albedo of about 0.07.

To account for Pluto's dimness, there was a strong tendency to suppose that it might be composed of rocky material and might not have an atmosphere. Even so, its mass could not be much greater than that of Earth if it were to be as dim as it was.

So, pretty soon, the astronomers began to divide the nine major planets of the Solar System into four gas giants, or "Jovian planets," and five small rocky worlds, or "Terrestrial planets." The Terrestrial planets were Mercury, Venus, Earth, Mars, *and* Pluto.

What a Terrestrial planet was doing way out there at the outer edge of the planetary system, when all the others hugged the Sun, could not be explained, but it was necessary to classify Pluto as such to allow for its dimness.

Still, even though Pluto had shrunk in size drastically at the moment of its discovery, it might still be the fifth largest planetary object in the Solar System, after the four gas giants, *if* it were slightly larger than Earth.

But is Pluto the size of Earth? In some ways, it bears the stigmata of what might be considered a very small planet. After all, some idea of a planet's size might be derived by studying its orbit.

The orbits of the planets are not, on the whole, very elliptical. The eccentricities of most of the planets are 0.05 or less. The eccentricity of Earth's orbit, for instance, is 0.017. This means that, to the unaided eye, most planetary orbits look just about circular.

The exceptions are the two smallest planets. Mars, with a mass only a tenth that of Earth, has an eccentricity of nearly 0.1. Mercury, with a mass only about a twentieth of the Earth's (half that of Mars) has an eccentricity of 0.2.

If we are going to associate low mass with high eccentricity, what do we do about Pluto? As its motion across the sky was studied for longer and longer periods, the details of its orbit were worked out, and the orbital eccentricity was found to be 0.25, higher than that of Mercury, and, in fact, the highest of any of the nine planets.

Does that mean that Pluto is even less massive than Mercury? Not necessarily. There is no *compelling* reason to associate low mass with high eccentricity. Thus, Neptune has little more than a twentieth the mass of Jupiter, and yet Neptune's orbital eccentricity is not greater than that of Jupiter, but is considerably less, only about a fifth that of Jupiter, in fact. Therefore, Pluto's high eccentricity may not, in itself, be a sufficient argument in favor of its being a very small planet—but it does give one pause to think.

The high eccentricity of Pluto's orbit means, by the way, that its distance from the Sun varies enormously in the course of its passage about that body. At its closest ("perihelion"), Pluto is 4,425 million kilometers (2,750 million miles) from the Sun. At the other end of its orbit, which it reaches a century and a quarter after perihelion, when it is farthest ("aphelion"), it is 7,375 million kilometers (4,583 million miles) from the Sun. This is a difference of 2,950 million kilometers (1,833 million miles).

This would make little difference to an exploring party on Pluto, of course. The Sun would be nothing more than a very bright star in Pluto's sky, and if it were a bit dimmer at aphelion than at perihelion, probably no one, but the expeditionary astronomer, would either know or care.

Its orbital eccentricity places Pluto, at times, a bit closer to the Sun than Neptune ever gets. At Neptune's perihelion, it is 4,458 million kilometers (2,771 million miles) from the Sun, while Pluto, at perihelion, gets 33 million kilometers (20.5 million miles) closer.

As it happens, in 1979, Pluto, approaching its perihelion, moved closer to the Sun than Neptune can be, and it ceased to be the farthest planet for a time. In each of Pluto's trips about the Sun it remains closer than Neptune for a twenty-year interval. This time, Pluto will reach its perihelion in 1989, and move out farther than Neptune again in 1999. It will not repeat this curious phenomenon until the years 2227 to 2247.

Another aspect of a planetary orbit is its "inclination," the amount by which it is tipped to the plane of Earth's orbit. Generally, the inclination of the planets is small. They circle in so nearly the same plane that if you make a small enough three-dimensional representation of the planetary system as far out as Neptune, it would all fit comfortably into one of those boxes that holds pizzas.

Once more, the smallest planet is a bit of an exception. While inclinations are usually 3 degrees or less, Mercury's is 7 degrees. If a high orbital inclination implies a small mass, what are we to make of Pluto's orbit, which has an inclination of some 17 degrees? Still, Uranus is considerably less massive than Saturn, yet Uranus also has a smaller inclination than Saturn has. We see, then, that inclination and mass don't have a necessary connection. Pluto's high inclination may not be significant, therefore—but again we are forced to think.

Pluto's high inclination means that although it seems to cross Neptune's orbit in a two-dimensional diagram of the planetary system, there is no chance at all of a collision between the two planets in the foreseeable future. Three-dimensionally, Pluto's high inclination carries it below Neptune's orbit so that the two planets are never separated by less than 1,300 million kilometers (800 million miles) at those times when the orbits seem to cross. Indeed, Pluto can, at times, be a bit closer to Uranus than it ever gets to Neptune.

The dimness of Pluto, which tells us it is smaller than originally expected, also tells us something else, because its reflected light is not constant.

If Pluto is a rocky planet, it may be that different portions of its surface reflect light with different efficiencies. There may be lighter rocks in one place than in another, or some rocks are frost-covered and others aren't. If this is so, then as the planet turns, its brightness should vary a bit. On the whole, there should be an overall variation with a period equal to its rotation.

In 1954, a Canadian astronomer, Robert H. Hardie, and a coworker, Merle Walker, measured the brightness very precisely and decided that Pluto rotates once every 6.4 days. (The best present-day figure is 6 days, 9 hours, and 18 minutes, or 6.39 days.)

This, too, puts Pluto's size in question. On the whole, it seems that the larger a planet is, the faster it rotates on its axis. Jupiter, the most massive planet, rotates in 9 hours 50 minutes; while Saturn, the second most massive, revolves in 10 hours, 14 minutes; and Uranus, the least massive of the gas giants, rotates in 17 hours, 15 minutes.

The Terrestrial planets, smaller than the gas giants, have longer rotational periods. Earth's is 24 hours, and the smaller orb of Mars rotates in 24 hours, 37 minutes. Mercury and Venus rotate very slowly indeed, but the Sun's tidal influences have something to do with that.

Yet Pluto, which can't possibly experience perceptible tidal influences from the very distant Sun, has a rotational period of over six days, which seems to be the mark of a very small planet. Again, this may be a coincidence, but we now have three characteristics—orbital eccentricity, orbital inclination, and rotational period—all of which seem to mark Pluto as very small. How far can coincidence stretch?

What is needed is a direct measurement of Pluto's diameter, but how is that to be done? At Pluto's huge distance, and rather small size, it looks like a mere dot of light even in a good telescope, even though it was fairly near perihelion at the time of discovery. (Had it been near aphelion, with its apparent diameter only three-fifths that at perihelion, the case would have been considerably more difficult.)

In 1950, however, the Dutch-American astronomer Gerard Peter Kuiper (1905–73) tackled the task, making use of the then-new 200-inch Palomar telescope. He turned it on Pluto and tried to estimate the width of the dot of light. It wasn't easy because the tiny orb of Pluto twinkles a bit and magnifying its size by telescope also magnifies the twinkles. The best Kuiper could do was to estimate that its size was 0.23 seconds of arc. (By comparison, Neptune's orb is never seen as less than 2.2 seconds of arc. Pluto's apparent width, then, is about one-tenth that of Neptune.)

An apparent width of 0.23 seconds of arc, allowing for Pluto's distance, would mean that its diameter would be something like 6,100 kilometers (3,800 miles). This would make our incredible shrinking planet considerably smaller than Earth. It would make Pluto, indeed, somewhat smaller than Mars, which has a diameter of 6,790 kilometers (4,220 miles). Instead of being the fifth largest planetary body, Pluto would then be the eighth largest, with only Mercury, among the major planets, smaller.

Not everyone accepted Kuiper's figure. The method of determining Pluto's diameter by looking at it through a telescope was simply too uncertain. There is, however, another way.

Every once in a while, Pluto, as it moves slowly across the sky, passes near a dim star. If it happens to move directly in front of the sky (an "occultation"), the star will wink out for a period of time. The time varies according to whether the star passes behind Pluto near one end of its orb or across its center. If we can get the exact position of the star and of the center of Pluto's orb, and if the minimum distance between the two can be determined and the time of the winking out of the star measured, then the diameter of Pluto can be worked out with pretty fair accuracy.

Of course it may be that Pluto might narrowly miss the star. In that case if one measured the distance between Pluto's center and the star, one can estimate the maximum diameter of Pluto, the one which would make it just possible to miss the star.

On April 28, 1965, Pluto was moving toward a dim star in the constellation of Leo. If Pluto were as large as the Earth, or even as large as Mars, it would occult the star, but it *missed.* From the fact that it missed, it could be calculated that the diameter of Pluto could not be more than 5,790 kilometers (3,600 miles) and might be substantially less.

So now it seemed that our incredible shrinking planet had to be no more than halfway between Mars and

Mercury in size. Its mass could not be more than one-sixteenth that of Earth—and it might be less than that.

The problem was finally solved, quite unexpectedly, in June 1978. The astronomer James Christie, working in Washington, D.C., was studying excellent photographs of Pluto taken by a 61-inch telescope in Arizona at high altitudes, where the interfering influences of the atmosphere were much reduced.

Christie studied the photographs under strong magnification, and it seemed to him that there was a bump on Pluto. Could it be that the telescope had moved very slightly while the photograph was being taken? No, for in that case all the stars in the field would have appeared as short lines; and they were all perfect points.

Christie looked at other photographs under magnification and they all had the bump. What's more, Christie noticed that the bump wasn't in the same place from picture to picture. In great excitement, Christie got still earlier photographs of Pluto, some as much as eight years old, and it became clear that the bump was moving about Pluto with a period of 6.4 days—Pluto's rotational period.

Either there was a huge mountain on Pluto, or else Pluto had a nearby satellite. Christie was sure it was a satellite and this was definitely proved in 1980, when a French astronomer, Antoine Labeyrie, working on top of Mauna Kea in Hawaii, made use of the technique of speckle interferometry. This showed Pluto as a pattern of dots, but it produced two such patterns, a larger and a smaller, with no connection between them. Pluto definitely had a satellite.

Christie named the satellite Charon (KAY-ron) after the name of the ferryboat pilot who, in the Greek myths, carried the shades of the dead across the River Styx into the underground kingdom of Pluto. (I would have chosen the name Persephone for the satellite, after the wife of Pluto, but Christie was apparently influenced by the fact that *his* wife was named Charlene.)

In 1980, Pluto passed close to another star. Pluto missed the star (at least as seen from Earth) but Charon passed in front of it and this occultation was viewed from an observatory in South Africa by an astronomer named A. R. Walker. The star winked out for 50 seconds, which gave Charon a minimum diameter of 1,170 kilometers (730 miles).

However, there was now a better way of determining size. Once you have a satellite, know its distance from the planet it circles, and the time it takes for one revolution, you can calculate the mass of the planet plus satellite. From the relative sizes of the planet and the satellite, assuming them to be of similar chemical composition, you can get the mass of each.

It turned out that Charon was 19,400 kilometers (12,000 miles) from Pluto. This is only one-twentieth the distance of the Moon from the Earth, so it's no wonder that, at Pluto's distance from us, so close a satellite went unnoticed for nearly half a century.

The mass of Pluto was calculated to be about 0.0021 (1/500) the mass of the Earth, so that the incredible shrinking planet turned out to be less massive than Mercury. In fact, it is only a little over one-sixth the mass of our Moon. All the criteria that seemed to show that Pluto was a very small planet were correct, after all.

As for Charon, it is about one-tenth the mass of Pluto.

Now that we know how small Pluto is, we can no longer imagine that it is made of rock. Given its size, it would not reflect enough light from bare rock to be as bright as it is. It must be an icy body, which would make it of lower density and of larger size, and which would allow it to reflect more of the sunlight that falls upon it.

It is now estimated that Pluto is about 3,000 kilometers (1,8 0 miles) across, a diameter about 7/8 that of our Moon, while Charon is about 1,200 kilometers (750 miles) across—just about the estimate obtained in the 1980 occultation.

This means that, in addition to the eight planets, there are seven satellites (the Moon, Io, Europa, Ganymede, Callisto, Titan, and Triton) that are more massive than Pluto. Pluto is neither the fifth largest planetary object in the Solar System, nor the eighth largest, but has shrunk to the sixteenth largest.

In the past some astronomers had tried to deny the apparently small size of Pluto in order to keep it as a massive and gravitationally significant body by suggesting that it had a smooth and icy surface and the dot of light we saw was not Pluto itself but the small reflection of the Sun on that polished surface. Others admitted the small size but tried to keep a high mass by imagining an enormous density.

Now, however, all tricks were done. Pluto was known to be tiny and its density could be calculated from its volume and its mass. The density turned out to be low, lower than anyone had expected (one more surprise). Pluto is only about 0.55 times as dense as water, less dense even than Saturn, which, at 0.7 times the density of water, had till then been the least dense known planetary object.

Pluto is too small to be made up of the gases hydrogen, helium, and neon, so it must be icy. Of the common ices, frozen methane (a combination of carbon and hydrogen atoms) is the lightest, being about half as dense as water. It may be, then, that Pluto is largely frozen methane, and if so it may have a thin, supercold atmosphere of methane vapor. Even at Pluto's distance from the Sun, some methane would evaporate, and the vapor would be cold enough to cling even to Pluto's small surface gravity.

Now consider— The mass of Ganymede, Jupiter's largest satellite, is 0.1 thousandth of the mass of its planet. The mass of Titan, Saturn's largest satellite, is 0.25 thousandth of the mass of its planet. The mass of Triton, Neptune's largest satellite, is 1.3 thousandths of its planet. The mass of our Moon, Earth's satellite, is, however, 12.3 thousandths of the mass of its planet.

To put it another way, the Moon has 1.23 percent of the mass of the Earth, and no other satellite, before 1978, had anything like that proportion of mass. The Earth and the Moon were the nearest thing to a double planet that we knew of.

Then came Charon, which is just about 100 thousandths (one-tenth) the mass of Pluto. Compared to Pluto, Charon is eight times the size of the Moon as compared to Earth. The Pluto-Charon combination has thus succeeded to the title of "double planet."

One last thing—Pluto and Charon are gravitationally negligible. They cannot have any measurable effect on Uranus's orbit. Yet the anomalies in Uranus's orbit, and, presumably, in Neptune's, too, are still there.

What causes them? Planet X does. It may still be out there somewhere, and it should still be a gas giant as originally presumed. The accidental discovery of the minuscule Pluto has merely diverted our attention from the search, so let's keep on looking.

ADDENDUM

This is my twenty-fourth (two-dozenth?) collection of essays from *The Magazine of Fantasy and Science Fiction* and you have no idea how glad I am that, a) I have survived to put together so many, b) the magazine has tolerated my column for so long, and c) the kindly people at Doubleday have been equally tolerant.

However, there are difficulties. Once I write an essay on a particular subject, I become sensitive to further developments on that same subject. I am annoyed if I am made to be out of date, or if something is done which I disapprove of in connection with the subject, or—well, almost anything. And as I write more and more essays I have increasing occasion for sensitivity and life becomes a little uncomfortable.

Thus, in this collection, I include, as Chapter 9, an essay I call "The Incredible Shrinking Planet." It describes the discovery of Pluto and how, in stages, the measure of its diameter has yielded smaller and smaller figures. At the start, it was confidently expected to be a planet that was larger than Earth, but that turned out to be wrong—in fact, increasingly wrong. Through a variety of investigations it is now known to be smaller than the Moon.

In my essay, the final figure I gave for Pluto's diameter was 1,850 miles. That essay was written in the fall of 1986 and a report I saw in the spring of 1987 reduced that figure further to one of 1,600 miles. If the smaller figure is correct, then Pluto has a diameter only three-quarters that of the Moon. Furthermore, since Pluto is made of light, icy materials rather than heavy, rocky material as the Moon is, the mass of Pluto may be only one-sixth that of the Moon.

Now a number of astronomers, annoyed, perhaps, at Pluto's failure to have a respectable size, wish to demote its status. Pluto should not be called a planet, they say, but should be listed as an asteroid.

Since I have written on that subject I feel that I have a right to make my opinion heard, too, and I feel that the suggestion is a ridiculous one.

There are three types of bodies in the Solar System that can be classified with absolute precision:

1) The Sun, which is the only object in the Solar System large enough to experience a nuclear ignition at its center and therefore emits intense visible radiation and is a star.

2) Planets, which do not emit intense radiation and which move in orbits that circle the Sun.

3) Satellites, which do not emit intense radiation and which move in orbits that circle a planet.

There is no chance of confusing one for another. We are not tempted to call even the largest planet a Sun, and we have no trouble distinguishing between a planet and a satellite.

However, the group of planets is exceedingly diverse, and this was first borne in on astronomers in the first decade of the nineteenth century when four planets, considerably smaller than any of the others, were discovered which circled the Sun in orbits that lay between those of Mars and Jupiter. The largest of these bodies (and still the largest despite the discovery of thousands of additional such objects) is Ceres, and its diameter is just about 640 miles, as compared with 3,013 miles for Mercury, the smallest planet known at that time. Ceres probably isn't more than 1/200 the mass of Mercury.

The astronomer William Herschel suggested that these small planets be called "asteroids" (from the Greek term meaning "starlike") because, through the telescope, they appeared as mere dots of light like stars, instead of expanding into visible circles of light like the larger planets.

The best way of defining an asteroid, then, is to call it a small planet that circles the Sun in an orbit lying between those of Mars and Jupiter. Again, my own feeling is that "asteroid" is a rotten name. If we want to differentiate among planets by size, then we ought to speak of "major planets" and "minor planets." What's more, very small objects such as "meteoroids" and dust particles, the diameters of which can vary from yards down to tiny fractions of an inch, ought to be called "microplanets." I even think that comets ought to be called "cometary planets."

Well, then, if scientists were to accept this eminently intelligent suggestion and call all objects circling the Sun as one or another subdivision of "planet," what ought they to call Pluto—a major planet or a minor planet?

To begin with, through long custom dating back for thousands of years, Mercury has been considered a planet, and no one suggests even today that it is anything but a major planet, albeit it might be considered the smallest of them. Let us therefore say that any object that is at least as large and as massive as Mercury is a major planet.

There are three satellites that are roughly as large as, or even a bit larger than, Mercury. These are Ganymede, Callisto, and Titan. However, there can be no confusion here. Ganymede and Callisto circle Jupiter, while Titan circles Saturn, so that despite their size, they are universally thought of as satellites and not as planets. Furthermore, these large satellites are made of relatively light, icy materials, so that even the largest of them, Ganymede, is only half as massive as Mercury.—So far, so good.

If we now omit Pluto, then the largest asteroidal body remains Ceres. It is considerably larger than any other asteroid, inside or outside the asteroid belt. It is also considerably larger than any known comet. It would be fair enough to say that a minor planet (or asteroid) was any object that moved in an orbit about the Sun and was the size and mass of Ceres or less.

That leaves a considerable gap between Ceres and Mercury. Mercury has nearly 5 times the diameter of Ceres and perhaps as much as 200 times the mass. It is in this gap that Pluto falls. Pluto's diameter, if taken to be 1,600 miles, is 2.5 times that of Ceres, while Mercury's diameter is nearly 2 times that of Pluto's. Pluto, therefore, is close to midway between. As far as mass is concerned, Pluto may be about 16 times as massive as Ceres, but Mercury is perhaps 16 times as massive as Pluto. Again Pluto is midway between.

But in that case, do we call Pluto a major planet or a minor planet? Perhaps, if Pluto had consciousness, it might prefer being by far the largest of the minor planets rather than being by far the smallest of the major planets. (Julius Caesar is supposed to have said, "I would rather be first in a little Spanish village, than second in Rome.")

However, my own suggestion is that everything from

Mercury up be called a major planet; everything from Ceres down be called a minor planet; and everything between Mercury and Ceres be called a "mesoplanet" (from a Greek word for "intermediate"). At the moment, Pluto is the only mesoplanet known.

Doesn't that make sense?

10

The Minor Objects

I was having lunch with an editor at a diner in my neighborhood last month, and the manager approached and said that he hated to interrupt us but there was a gentleman who wanted to be introduced to me. I sighed, looked at my companion a bit nervously (I always have the feeling that I may be suspected of having arranged such matters in order to make an impression), and said, "Well, bring him here."

Up came a man of middle size, rather thin, with dark eyes, a prominent Adam's apple, a shirt open at the collar, and a day's growth of beard. I did not stand up, for among the few privileges of advancing years is that of keeping one's seat when a younger man would feel compelled to rise. After all, as the knee joints ripen, the wisdom of not loading them with unnecessary labor is evident to all.

I placed a pleasant smile on my face and said "Hello" to the newcomer, who said to me earnestly,

"Dr. Asimov, my name is Murray Abraham, and I want to tell you that your book on Shakespeare—"

That's as far as he got, for *now* I rose to my feet, and with explosive energy, too.

I said, with the utmost conviction, "You are *not* Murray Abraham. You are Antonio Salieri!"

After that, the conversation was simply chaos. I wouldn't listen to him tell me about my books. I wanted to tell him about his performance in the motion picture *Amadeus,* and since I was older than he was, I suppose he felt it a necessary courtesy to let me have my way, finally. I never found out what he wanted to say about my book on Shakespeare.

It was like this, you see. I very rarely go to a motion picture theater, chiefly because I am too busy at my typewriter or word processor. (You'd be surprised how being a prolific writer eats into your time.) I did, however, seize the occasion to see *Amadeus.*

I watched with awe as F. Murray Abraham (whom I had never seen before) played the difficult role of the partly villainous, partly pathetic Salieri. About halfway through the picture, I turned to my dear wife, Janet, and said, "That guy Abraham is going to win an Academy Award for this."

I hadn't seen any of the competition, but I was quite certain that no other motion picture portrayal that year could possibly come up to Abraham's. I knew perfection when I saw it.

Of course, Abraham *did* win the Academy Award, and I have never enjoyed such a victory more. I considered it as much a tribute to my own judgment as to Abraham.

That's why I was so excited when I met him, and

that's why I denied his name. Now and forever after, he may be F. Murray Abraham to himself, but he is Antonio Salieri to me.

And as I meditated on the difficulty of distinguishing between an actor and his role, a chain of thought led me to the matter of distinguishing between a comet and an asteroid. So here goes—

If we're going to distinguish between the two great classes of minor bodies of the Solar System, let's begin by defining each.

Asteroids are a swarm of small bodies that circle the Sun between the orbits of Mars and Jupiter. Some of them are fairly large, and one of them, Ceres, is about 1,000 kilometers in diameter. There are several dozen asteroids with diameters of over 100 kilometers, but the bulk of the possibly 100,000 asteroids that may exist are small objects, no more than a few kilometers across.

A second swarm of minor objects is thought to circle the Sun at a much greater distance. Whereas the asteroids orbit the Sun at distances of about 400 million kilometers, the second swarm may lie as far as one or two light-years away, and so are some 35,000 times as far from the Sun as the asteroids are. Let's call the bodies of this far-off second swarm "cometoids." (This term is my own invention and is not used by astronomers, as far as I know.)

Naturally, no astronomer has ever studied, or even seen, any of the cometoids circling the Sun in this far-off region. The cometoids out there are too distant and too small to be detected in any way. Their existence can only be inferred from the existence of comets, and from the close study of cometary orbits, structure, and behavior. That's why I call the bodies of the hypothesized distant swarm cometoids, as a back-formation from "comet."

Cometoids and asteroids are both relatively small, solid bodies in orbit about the Sun, but not only are the former much farther from the center of the Solar System, they are thought to be much more numerous as well. I have seen estimates to the effect that there may be as many as 100 billion cometoids, or about a million cometoids for every asteroid.

Differences in distance and number are, however, trivial. If those were the only differences there were, an individual cometoid and asteroid set side by side might be indistinguishable.

There is, however, an important difference in chemical structure that depends directly on the difference in distance.

Both cometoids and asteroids were presumably formed when the Solar System was taking shape. What's more, they were formed from the same vast cloud of dust and gas from which the Sun and the planets were formed. Astronomers are pretty certain this cloud was made up chiefly of hydrogen and helium, with an admixture of such other atoms as carbon, nitrogen, oxygen, neon, argon, silicon, and iron.

Hydrogen, helium, nitrogen, oxygen, neon, and argon are gases that do not easily solidify, even at great distance from the Sun. Hydrogen combines with oxygen, however, to form water, with nitrogen to form ammonia, and with carbon to form methane. These substances all freeze into solids that, in appearance, resemble ordinary ice (that is, frozen water), and so they are lumped together as "ices."

The remaining elements, making up somewhat less than half a percent of the total, collect into metals and rocky substances.

Out of this and other considerations, the American astronomer Fred Lawrence Whipple (1906–) advanced the notion, in 1950, that cometoids are "dirty snowballs"; large lumps of ice (chiefly water-ice) with stony and metallic particles distributed as dust and as occasional larger pieces. It is even conceivable that some cometoids might have a solid rock and metal core.

Some calculations would make it appear that a cometoid is two-thirds ice in mass, and one-third rock and metal.

Cometoids are dirty snowballs, however, only because they have been formed far from a Sun that was coming into being even as the cometoids themselves were being shaped. The young Sun poured heat in every direction and had a strong solar wind, too. The heat vaporized those substances most easily vaporized and the solar wind swept those vapors outward and away. Large objects such as Jupiter and Saturn could hold on to any vapors produced, thanks to their enormous gravitational field, but anything the size of minor bodies, such as the cometoids, couldn't. They couldn't hold on to hydrogen, helium, and neon, which remained vapors even under the feeble heat of the distant Sun. They *could,* however, hold on to those substances which solidified into ice at the low temperature of the distant reaches of space.

If minor bodies formed *near* the Sun, however, as,

for instance, in the asteroid belt, the results would be different.

The asteroids, forming in the comparative neighborhood of the Sun, would be sufficiently affected by the heat of the Sun for all the ice that might have formed to have eventually turned to vapor. In fact, the heat would be such that the ice might not collect in the first place. All this vapor would then be swept away by the solar wind and driven into the far reaches of the Solar System, where it might then contribute to the formation of the cometoids.

The asteroids, therefore, are formed almost entirely out of the rocky and metallic bits left over. It is this paucity of building material that may make the asteroids so much fewer than the cometoids and, on the whole, somewhat smaller, too.

To make a distinction, then, that is not trivial, asteroids are built up of rock or metal or a mixture of the two, with ice present in minor quantities or not at all. Cometoids are built up of ice, chiefly, with rock and metal forming a minor impurity.

An astronomer, viewing a small body at telescopic distance, could label it a cometoid or an asteroid according to its manner of reflecting light. An icy cometoid would reflect a much larger percentage of any light falling upon it than a rocky or metallic asteroid would.

Then, too, because of this difference in chemical composition, something happens to cometoids that never happens to asteroids.

Every once in a while, the distant cometoids are perturbed in their majestic million-year orbit about the

Sun. There might occasionally be a collision between two cometoids that would transfer energy from one to the other, slowing the former and speeding the latter. Or else the gravitational pulls of the nearer stars might, depending on the position of the stars, slow or speed a cometoid.

A cometoid that gains energy and speed moves farther from the Sun and may eventually be lost to the Sun forever, taking a virtually endless pathway through interstellar space. A cometoid that loses energy and speed moves nearer the Sun, and perhaps penetrates the regions in which the large planets exist.

The gravitational effect of the outer planets on a cometoid wandering through their neighborhood may force that cometoid into a radically new orbit—one that brings it into the near vicinity of the Sun at one end. Planetary influences may even trap it (or "tame" it) to the point where it remains within the planetary portion of the Solar System through its entire orbit. It then becomes a "short-term" comet. Instead of circling the Sun in millions of years, it circles it in no more than a century or so—or even less.

Cometoids do not long survive their close approaches to the Sun, or at least not long in astronomical terms. Whether cometoids undergo an increase in energy and move away from the Sun forever, or whether they lose energy and pass on to eventual destruction in the solar neighborhood, they are lost to the cometoid belt. It is estimated, however, that in the entire 4.5 billion years that the Solar System has existed, only about one fifth of the cometoid horde has been lost. By far the greater part remains.

Let's concentrate, now, on the cometoids that approach the Sun. When they do so for the first time, they are exposed to the Sun's heat as they never were when they were at home in the far-off swarm. As the cometoid is heated, the ice evaporates and the dust particles of rock and metal are freed. The gravitational pull of the cometoid is too weak to keep those dust particles firmly bound to the surface and the upward movement of vapor carries the particles along. The vapor and dust form a kind of atmosphere surrounding the cometoid and the dust particles glitter in the Sun. The cometoid, as it approaches the Sun, thus develops a hazily luminous "coma." The coma is swept back by the solar wind into a "tail."

The coma and tail grow larger and brighter as the cometoid approaches the Sun, until, if the cometoid is large enough and approaches Earth closely enough in its passage, it becomes a magnificent sight, with the tail arching a long distance across the sky. It is only in this form that we can see and study a cometoid. The hazy, tailed object into which a cometoid is converted is called a "comet." This is from the Greek word for "hair" because the tail, to the imaginative Greeks, resembled the long, unbound hair of a person, streaming backward as the comet moved across the sky.

The distinction between a *comet* (rather than a cometoid) and an asteroid is child's play.

An asteroid is just a luminous dot in the sky, even when seen by the best telescope. It looks just like a star (hence the name "asteroid," meaning "starlike") but is distinguished from stars by the fact that it moves against the background of the true stars.

A comet, however, is a much brighter object, hazy in appearance and irregular in shape. The large comets have long tails and are bright enough to be seen by the unaided eyes. Even small and distant comets, that can only be seen telescopically, will show a haze unless they're very far from the Sun.

And yet there is another difference between the two. Whereas an asteroid is a permanent object, a comet ages rapidly, and the distinction between an *old* comet and an asteroid can be blurred.

Every time a comet skids about the Sun, a considerable quantity of its substance is vaporized and swept away—and never returns. Each time the comet passes the Sun, then, it is smaller than the time before, and, eventually, it may vanish entirely.

Astronomers have watched this happen. The most celebrated case is that of Comet Biela, so called because its orbit was first worked out in 1826 by an Austrian amateur astronomer, Wilhelm von Biela (1782–1856). The comet had a small orbit and reached perihelion every 6.6 years. It was observed in 1846, when it was found to have lost enough material to produce a split. Instead of one comet, two appeared. In 1852, the double comet appeared again, the two fragments widely separated and the smaller fragment very faint.

After that, Comet Biela was never seen again. Apparently, it had been totally vaporized—or, to put it more dramatically, it had died. Other such splittings and disappearances have been seen since.

But cometary deaths may occur in different ways. That of Comet Biela, death by total evaporation, is most spectacular, but a comet can also die a quieter and much more prolonged death.

Some comets may well contain more solid dust mixed with its ice than others do and the dust may be distributed unevenly. Portions of the cometary surface that are particularly dusty would vaporize more slowly than those portions where the ice is purer. For that reason the cometary surface may melt away in such a fashion as to form plateaus of dusty areas interspersed with valleys where dust-poor areas had evaporated. Occasionally dust-rich plateaus may be undercut and collapse, exposing fresh surface to evaporation and causing a sudden temporary brightening of the comet. (Such brightenings are frequently observed.)

Such collapses help spread dust over the surface generally. In addition, some dust which is liberated by ice evaporation and which lifts off the surface may drift back as the comet recedes from the Sun. The dust is much more likely to do this than the ice vapors will.

As a comet ages, then, its surface gets more dusty. The dust eventually builds up into a thick layer that hides and insulates the ice from solar heating, so that an old comet forms very little haze and no tail.

The best example of an old comet is Comet Encke, so called because its orbit was first computed, in 1819, by a German astronomer, Johann Franz Encke (1791–1865). Comet Encke has the smallest orbit of any known comet and the shortest period. It reaches perihelion every 3.3 years. It has been closely observed dozens of times and each time it exhibits a faint haze—one that is just enough to identify the object as a comet.

Under such circumstances, a comet can last a long time, as it dribbles away its buried ice from under its protective layer of compacted dust. In the early stages, of course, a particularly thin portion of dustcover may burst from the pressures of heated ice beneath, and a gout of vapor and dust may emerge from the newly exposed ice. This, too, would cause cometary brightening. However, Comet Encke is past that stage.

Even an old comet must eventually give up all its ice, or at least reduce the dribble of vapor to so small a quantity that it can no longer be observed. It might even be that some comets may have small cores of rock and metal that persist after the ice is totally gone.

How, then, do you tell a *dead* comet (whether it has well-hidden ice or none at all) from an asteroid?

One difference that remains is the nature of the orbit. Almost all asteroids have orbits that are entirely between those of Mars and Jupiter. What's more, those orbits are not very eccentric or very far inclined to the "ecliptic" (that is, to the plane of Earth's orbit).

Cometary orbits, on the other hand, are characteristically very eccentric and commonly have a high inclination to the ecliptic.

If, then, we were to discover asteroids with orbits that show high eccentricity and inclinations, we might wonder whether we really have an asteroid—or a dead comet.

There are such suspect asteroids with orbits that bring them close to the Sun periodically, so that they have perihelia that are nearer the Sun than the planet Venus is. These are the "Apollo objects," and, of them, the most spectacular, until recently, has been Icarus, an asteroid discovered in 1948 by the German-American astronomer Walter Baade (1893–1960). It was the fifteen hundred sixty-sixth asteroid to have its orbit determined, so it is officially known as "1566 Icarus."

At its perihelion, Icarus is only 28.5 million kilometers from the Sun. The planet Mercury at perihelion is 45.9 million kilometers from the Sun, so that Icarus reaches a distance from the Sun only three-fifths that of Mercury's best mark. The asteroid has been aptly named, then, after the character in Greek mythology who flew with his father on homemade wings. Icarus, in his arrogance, flew too near the Sun, so that the wax holding the feathers of his wings to its wooden framework melted. Off came the feathers and down fell Icarus to his death.

At its aphelion, Icarus is at a distance of 300 million kilometers, well within the asteroid belt. Its eccentricity (the measure of the elongation of its orbit) is 0.827, the highest then known for an asteroid. Its inclination is also quite large, being 23 degrees. It isn't too unreasonable, therefore, to wonder if Icarus might be a dead comet.

Then, on October 11, 1983, the Infra-Red Astronomical Satellite (IRAS) detected an asteroid with an unusually rapid apparent motion against the stars. (This rapid motion at once showed it to be near Earth and made it seem very likely that it was an Apollo object.)

The asteroid was first called 1983 TB, according to a system used to identify asteroid sightings. The IRAS sighting didn't give much information about the asteroid but it gave enough to allow it to be tracked down by ordinary telescopes. Its orbit was then worked out. Since it was the thirty-two hundredth asteroid to have its orbit determined, it might be called "Asteroid

3200.'' (Just about as many new orbits have been worked out since 1948, you'll notice, as in all the years before 1948—something that must be attributed to the coming of computers.)

The remarkable thing about Asteroid 3200 is that at perihelion it is closer to the Sun even than Icarus. The perihelion distance of Asteroid 3200 is 21 million kilometers; only three-quarters that of Icarus, less than half that of Mercury, and one-seventh that of Earth. The asteroid was promptly named Phaethon, after a character in Greek mythology, the son of the Sun-god, who persuaded his father, Helios, to let him take the reins of the solar chariot for a day. With Phaethon's unskillful hands at the reins, the solar horses careered madly across the heaven. Lest he destroy the Earth, Phaethon was struck dead by a thunderbolt from Zeus. Phaethon clearly approached the Sun even more closely than Icarus did in the myths—and in astronomy.

At aphelion, ''3200 Phaethon,'' as it should now be called, is about 385 million kilometers from the Sun, considerably farther out than is true of Icarus. With Phaethon's perihelion closer and its aphelion farther than is the case for Icarus, you can see that Phaethon's orbit is even more elongated than that of Icarus, and its eccentricity is higher. Phaethon's eccentricity is 0.890, a new high for an asteroid. In inclination, Phaethon is 22 degrees compared to Icarus's 23. Phaethon returns to perihelion every 1.43 years (522 days), while Icarus does so every 1.12 years (409 days).

Well, then, is Phaethon a dead comet?

When Phaethon was first observed by ordinary telescope, it was quite far away and receding. Astronomers watched for its next approach to see if, under the most favorable conditions, a dribble of vapor and dust could be seen. In December 1984, it passed near Earth, and no trace of coma could be seen. In fact, it looked like a stony asteroid, so that if it were a dead comet, it was a *very* dead comet.

Does any way remain of making a distinction between a completely dead comet and an asteroid that was never a comet at all? Oddly enough, there is—after a fashion.

As comets age, the dust that is liberated as part of the coma and tail continues to move about the Sun in the cometary orbit. Little by little, for a variety of reasons, the dust particles are distributed throughout the orbit, though a heavier concentration may remain, for a period of time at least, in the neighborhood of the comet, or in the place where it used to be if it has died by total evaporation.

Every once in a while, the Earth, in its orbit, cuts across such a dust swarm and the particles heat and vaporize in the atmosphere, forming meteoric streaks at a greater rate than is customary on ordinary nights. Once in a long while, in fact, Earth cuts across a heavy concentration of such particles and the result resembles the falling of luminous snowflakes (though none survive to reach the ground). There was such a major meteor shower over the eastern United States on the night of November 12, 1833, and this incident began the serious study of meteoritics.

There are a number of such "meteor streams," as they are now called. Their orbits have been worked out and are found to be cometlike in character. Sometimes the particular comet that is associated with them is alive and can be identified. One meteor stream has been found to follow the orbit of vanished Comet Biela and when its particles enter Earth's atmosphere, they are called "Bielids," in consequence.

If an Apollo object is a dead comet, might it not be possible that a meteor stream occupies its orbit? It would seem so unless the comet has been dead too long, for, with time, the dust particles are gathered in by the planets and satellites they pass, or are dispersed through space in some fashion.

As it happens, most Apollo objects have not yet been found to be accompanied by meteor streams, although two of them, 2101 Adonis and 2201 Olijato, have orbits that are at least close to the known orbits of two such streams.

Fred Whipple pointed out, however, that Phaethon's orbit is very close to that of a well-known meteor stream known as the "Geminids." The orbits are so nearly identical that it is difficult to suppose it to be coincidence only. Therefore, if any of the Apollo objects is really a dead comet, Phaethon is.

As in the case of all Apollo objects, the question arises as to whether Phaethon might ever strike the Earth. If so, it would be a horrible catastrophe, for Phaethon is estimated to be nearly five kilometers in diameter. Fortunately, Phaethon crosses the ecliptic at a point well inside Earth's orbit so that it remains several million kilometers away at even its closest point to Earth.

However, the gravitational pulls on Phaethon by the various planets combine to move the point of crossing the ecliptic farther from the Sun. If this continues, then, according to some calculations, in two hundred fifty years the orbits will actually intersect and there would then be a small chance that both Earth and Phaethon would arrive at the orbital intersection point simultaneously before the ecliptical crossing point moved still farther outward and made collision impossible again.

On the other hand, as Phaethon makes closer and closer approaches to Earth, Earth's gravity will send it into a new orbit less dangerous to Earth. An actual collision is *most* unlikely.

[NOTE: *The probe Giotto studied Comet Halley at close range several months after this essay was written. It supported the notion of a dust covering. Indeed, Comet Halley was coal-black, thanks to a thick surface layer of dust.]*

Part III
Beyond the Solar System

11

New Stars

Any one of us can make up a personal "Book of Records," if we wish to. What is the longest time you have ever gone without sleeping? What is the best meal you ever had? What is the funniest joke you ever heard?

I'm not sure that the effort involved is worth it, but I can tell you, very easily, the greatest astronomical spectacle I ever witnessed.

Living in the big cities of the Northeast, as I do, there isn't much in the way of astronomical spectacles I can see. Between dust and artificial light, I'm lucky if I can occasionally make out the Big Dipper in the New York night sky.

And yet, back in 1925, there was a total eclipse of the Sun which was visible from New York City—just barely. It was called the Ninety-sixth Street Eclipse because north of Ninety-sixth Street in Manhattan, totality was not quite achieved.

I lived about ten miles to the south of that limiting line, however, so I was all right, for totality endured in my neighborhood for a short time. The trouble is, however, that I was only five years old then, and can't, for the life of me, remember whether I saw the eclipse or not. I *think* I remember seeing it, but I may be only kidding myself.

Then, in 1932 (in August, I think) there was an eclipse visible in New York City that was about 95 percent total. It was a thrilling time, for the Sun was reduced to a thin crescent and everyone stood about in the street and, even more so, on rooftops to watch it. (I think that most people chose rooftops to be closer to the Sun and get a better view.) We all stared through smoked glass and exposed photographic film, which were quite inadequate to the task, and why we didn't all go blind, I don't know. In any case, *that* eclipse I saw. I was twelve years old and I remember it well.

But then, on June 30, 1973, I was on the ship *Canberra* off the coast of West Africa and I saw a total eclipse of the Sun, beautifully. It lasted five minutes, and what impressed me most was its ending. A tiny dot of bright light appeared, and suddenly spread out in half a second to become too bright to look at without filters. It was the Sun coming back with a roar—and that was the most magnificent astronomical spectacle I ever saw.

There are other spectacles in the sky, which may not be so spectacular as a total eclipse of the Sun, but which are more interesting to astronomers—and even to us, once we understand them thoroughly. There are, for instance, such things as apparently new stars. A solar eclipse is only a case of the Moon getting in front of the Sun, and this is a regular phenomenon that is easily predictable for centuries ahead. New stars, on the other hand, are—

But let me begin at the beginning.

In our Western culture, it was taken for granted for a long time that the heavens were changeless and perfect. For one thing, the Greek philosopher Aristotle (384–322 B.C.) said so and for eighteen centuries it was hard to find anyone willing to argue with Aristotle.

And why did Aristotle say that the heavens were changeless and perfect? For the best possible reason. It seemed so to his eyes, and seeing is believing.

To be sure, the Sun shifted position against the stars, and so did the Moon (which went through a cycle of phases as well). Five bright starlike objects, which we nowadays call Mercury, Venus, Mars, Jupiter, and Saturn also shifted position, in more complicated fashion than the Sun and Moon did. However, all these motions, together with other changes involving phases and brightness, were quite regular and could be predicted. In fact, they were predicted by methods that were slowly improved by astronomers, starting with those bright people, the Sumerians, about 2000 B.C.

As for changes that were irregular and unpredictable, Aristotle maintained that these were phenomena of the atmosphere, and not of the heavens. Examples of such things are clouds, storms, meteors, and comets. (Comets, Aristotle thought, were just burning gases high in the air—lofty will-o'-the-wisps.)

Aristotle's notion of changeless perfection fitted in nicely with Judeo-Christian ideas. According to the Bible, God made the Universe in six days and then rested on the seventh because, presumably, there was no more left to do. It seemed blasphemous to suppose that God would suddenly realize that he had left something out and get into the creation business again, after the six days were long over, in order to create a new star or, for that matter, a new species of life.

To be sure, the Bible describes God as endlessly interfering with human beings and becoming wrathful at the least little thing and sending down Flood and Plagues, and ordering Samuel to wipe out the Amalekites, including the women, children, and cattle, but that was only because human beings seemed to irritate him. He left the stars and the species alone.

So what with Aristotle and Genesis, people of our Western tradition, if they saw a new star in the sky, would probably have averted their eyes nervously and decided they shouldn't have taken that last swig of mead or sack or whatever they had been tippling.

Besides, they weren't likely to pay attention to a new star even if it appeared. Few people looked at the sky with any intentness or bothered to memorize the patterns of stars and remember this combination here and that combination there. (Do you?) Even astronomers, who watched the heavens professionally, were chiefly interested in the peregrinations of those heavenly bodies ("planets") that moved with reference to the others: the Sun, Moon, Mercury, Venus, Mars, Jupiter, and Saturn. From those motions they developed the pseudoscience of astrology, which still impresses unsophisticated people (i.e., the majority of humanity) today.

As for the other stars, which retain their positions relative to each other, one might note the Big Dipper and the Square of Pegasus and other simple configurations of relatively bright stars, but no more than that. Therefore, if a new star appeared and produced a change in some unnoted pattern, the chances are that it would go unnoticed except by a very few, and those few would not be able to convince others that it was *really* a new star. I can hear the conversation now:

"Hey, look, that's a new star!"

"Where?—What makes you think that's a new star?"

"It wasn't there last night."

"You're crazy."

"No. Honest. Cross my heart. Hope I die. That's a new star."

"So? Even if it is, who cares?"

Of course, if a new star appeared and was really bright, it might be noticed. The brightest star in the sky is Sirius, but brighter still are several of the planets, including Jupiter and Venus. If a new star were of "planetary brightness"—that is, if it rivaled the planets in brightness and was brighter than any ordinary star—it would be hard to ignore.

The first account of the sighting of such a new star involves Hipparchus (190–120 B.C.), a Greek astronomer who worked on the island of Rhodes. Unfortunately, none of his writings have survived, but we know enough from the writings of later scholars to be able to judge that he was the greatest astronomer of antiquity.

The oldest reference to his sighting of a new star that still survives today is in the writings of the Roman encyclopedist Pliny (23–79 A.D.), who wrote two centuries after Hipparchus. He states that Hipparchus had spotted a new star and had thereupon been inspired to prepare a map of the stars in the sky.

That sounds reasonable to me. Hipparchus must have studied the visible night sky as few others ever did, and he could therefore recognize a particular star as being a new one where others would not. What's more, he might well have wondered if other such new stars had shown up earlier and had evaded his notice. If he prepared a map, then any star which looked even vaguely suspicious could be compared with that map and be revealed as a new star (or as an old one) at once.

Despite Hipparchus's map, and its improvement by another Greek astronomer, Claudius Ptolemy (100–170), three centuries later, no new stars were definitely spotted by Western observers for seventeen centuries after Hipparchus. You have to give credit to Aristotle and Genesis for that.

However, there was one civilization on Earth that was advanced in science and that had never heard of either Aristotle or Genesis until 1500 or thereabouts. That was China. Unhampered by religious views concerning the nature of the heavens, they were quite ready to see any new stars that might appear in the sky. (They called them "guest stars.")

The Chinese reported five particularly bright new stars, each one of which remained visible for six months or more. (In other words, these were not only new stars, appearing in the heavens in a spot where no star had earlier been seen, but they were *temporary* as well, for they eventually disappeared, whereas ordinary stars remained in place, apparently, forever.)

For instance, they reported a very bright new star in the constellation of Centaurus in 183. (Of course, they had their own names for various star groupings, but we are able to translate their constellations into ours.) According to the Chinese, the new star, at its peak, was of planetary brightness, brighter than Venus actually, and stayed visible for a year. However, it was far in the southern sky and was not visible from most of Europe. It would have been visible from Alexandria, which was then the center of Greek science, but Alexandria was past its best days, and the last Greek astronomer of note, Claudius Ptolemy, was dead.

The next bright new star appeared in Scorpio in 393, but it was less bright than the one in Centaurus, and not quite of planetary brightness. It was as bright as Sirius (the brightest ordinary star) for a short time and remained visible for eight months. There were no reports from Europe, however. The Roman Empire had turned Christian, and such scholars as existed were debating theology rather than the details of the sky.

About six centuries passed before another new star of planetary brightness was reported by the Chinese. This was in the constellation Lupus, again far in the southern sky, and it appeared in 1006. It was the brightest star ever reported by them, and may well have been the brightest star to have appeared in the sky in historic times.

According to some modern astronomers working with the Chinese reports, it must, at its very brightest, have been 200 times as bright as Venus ever gets, which means that it was perhaps one-tenth as bright as the full Moon. (Since it was just a dot of light, all that brightness packed into it must have made it rather dazzling to look at.) It only remained at or near maximum brightness for a matter of a few weeks, but it faded slowly and didn't sink to invisibility for about three years.

The Arabs, who had made good use of the Greek

heritage of science, and were the foremost astronomers in the West at this time, also reported it. Only a couple of very dubious reports, however, have been dug out of European chronicles that might apply to the star, but, then, Europe was just emerging from the Dark Age.

Then, in 1054 (on July 4, of all days, according to some reports), a very bright new star blazed out in the constellation of Taurus. It was not quite as bright as the new star in Lupus half a century earlier, but at its peak it was two or three times brighter than Venus.

For three weeks it remained bright enough to be seen in daylight (if one knew where to look) and it even cast a dim shadow (as Venus sometimes does) at night. It stayed visible to the naked eye for nearly two years and it was the brightest new star to be seen in historic times that was far enough up in the north sky to be easily visible from Europe. It was even in the Zodiac, the region of the sky most studied by astronomers of the day.

There are Chinese and Japanese reports of the new star of 1054, but in the West, despite the fact that it was high in the sky and in the Zodiac, to boot, there was almost nothing. In recent years, an Arab reference has been uncovered that might concern the new star, and even an Italian reference has been uncovered, but certainly these are minor considering the spectacular blaze it must have created in the sky.

Finally, in 1181, a new star appeared in Cassiopeia, again high in the northern sky. It didn't become very bright, however, and was not even as bright as Sirius. Though it was reported by the Chinese and Japanese, again it went unnoticed in Europe.

That's five such new stars in the space of a thousand years that were faithfully reported by the Chinese yet went all but unnoticed in the West. All the West had was Pliny's tale of Hipparchus's sighting and that presented so little detail (and Pliny's ability to believe anything at all, however ridiculous, was so notorious) that it might have been considered as legendary, at best.

Let me, however, make mention of still another ancient new star that must have been even more spectacular than the five of Chinese reports and the sixth of Hipparchus.

In 1939, the Russian-American astronomer Otto Struve (1897–1963) discovered a faint trace of nebulosity in the southern constellation of Vela. Between 1950 and 1952, this was followed up by the Australian astronomer Colin S. Gum (1924–1960), who published his findings in 1955. He was able to show that it was a large cloud of dust and gas that filled one sixteenth of the sky and it is called the "Gum Nebula" in his honor.

Astronomers now know that this sort of dust and gas cloud is a sign that a new star had once appeared at its center. The center of the cloud is just about 1,500 light-years from us and this is considerably closer than any of the new stars reported in ancient times. (Naturally, none of the ancient observers had any notion as to how far away the new stars—or any stars—were, but astronomers have ways of estimating those things now.)

Since the new star of the Gum Nebula was much closer than the others, it must also have been much brighter. Astronomers now feel that at the peak of its brightness it may have been as bright as the Full Moon. To anyone watching it, it must have appeared to be a small bit of the Sun which had broken off and been stuck immovably in the sky.

No one who saw *that* new star's blaze could possibly have failed to notice it, and yet there is no report of it from anywhere. Given that it was far to the south, it might seem unbelievable that it left no impression whatever.

But there's no mystery to it. From the size of the Gum Nebula and the rate at which it is enlarging, you can tell when the whole thing was the size of a star—and that was 30,000 years ago, in the Old Stone Age. It was noticed, I'm sure, but no record could be kept.

Too bad. That astronomical phenomenon must have been something to see, except that over a period of a few weeks it would probably have been impossible to look at the star except through smoked glass or through a veil of clouds.

Let's consider what happened after 1181, when the fifth and last bright new star appeared that we only know of from Chinese records.

Nearly four hundred years passed before the next new star appeared and by then things had changed in Europe. The continent had progressed and was advancing rapidly in science and technology. The Polish astronomer Nicolaus Copernicus (1473–1543) had published a book on astronomy in 1543 that advanced the theory that the Sun, not the Earth, was the center of the planetary system, and that the Earth was itself a planet, like the other planets. This began what we now call the "Scientific Revolution."

Three years after Copernicus's book was published,

Tycho Brahe (1546–1601) was born in the southernmost province of Sweden, which was then part of Denmark. He turned out to be the best astronomer since Hipparchus.

The year is 1572. At that time Europeans still had no notion that any new star had ever appeared in the sky, if one discounted Pliny's possibly fanciful tale about Hipparchus. In that year, Tycho (he is usually known by his first name, as was true of many of the scholars and artists of the time, especially in Italy) was only twenty-six years old and still unknown.

On November 11, 1572, Tycho, leaving his uncle's chemical laboratory, was thunderstruck to see a new star in the sky. He couldn't miss it. It was high in the sky, and it was in the very well-known constellation of Cassiopeia. Cassiopeia consists of a lopsided *W* built up of five fairly bright stars, and that *W* is almost as familiar a combination as is the Big Dipper. But now the Cassiopeia *W* consisted of *six* stars, the sixth star, a little to one side of the *W*, being far brighter than all the rest combined. It was actually brighter than Venus but it couldn't be Venus, because that planet is never found in that part of the sky.

Tycho asked each person he met if he could see the star, for, under the conditions, he dared not trust his eyesight. (They all saw it.) He also tried to find out if it had been in the sky the night before, for he hadn't had occasion to look at the sky for some time, but, of course, no one could say.

Actually, there seems to have been a report from a German astronomer, Wolfgang Schuler, that might indicate that he saw the star five nights earlier than Tycho did. However, Schuler didn't follow up the matter, and Tycho did. Tycho began a series of nightly observations with excellent instruments he devised himself.

Tycho's new star was quite near the celestial north pole, so it never set and Tycho could observe it day and night, for (to his surprise) it was bright enough to be seen in the daytime—at least when he first observed it. Even though it slowly faded from night to night, it was a full year and a half before it faded entirely from sight.

Tycho wondered what to make of the new star, which, as far as he knew, was the *only* new star that had ever appeared in the heavens, if one discounted the vague Plinian reference to Hipparchus.

Since it certainly represented a change in the heavens, it should, according to Aristotle, be an atmospheric phenomenon. If it were, it should be closer to Earth than the Moon.

Now, if the Moon's position against the stars is carefully noted, at a given time, from two points on Earth that are separated by a reasonably long distance, the Moon seems to be in a slightly different position with respect to nearby stars as seen from each point. This is the "parallax" of the Moon, and if the size of the change in position, and the distance between the two points of sighting are known, the distance of the Moon can be calculated by trigonometry. It wasn't easy to do this in the days before there were accurate clocks and easy communication between different points on Earth, but it had been managed and it was known that the Moon was about a quarter of a million miles from Earth.

The distance of no other heavenly body was known because no object in the sky other than the Moon gave a measurable parallax. Since the distance of an object varies inversely with its parallax, that meant that all visible objects that were not atmospheric phenomena were farther than the Moon. Or you might put it this way: once you left the Earth's atmosphere, the first object you came to on your journey away from Earth was the Moon. Even the ancient Greeks were certain of that.

If, then, Tycho's new star were atmospheric and were closer to us than the Moon is, that new star should have an even larger parallax than the Moon and its parallax should be even easier to measure.

Not so. All of Tycho's efforts went for naught. The new star showed no parallax at all; that is, its parallax was too small to measure. This meant that the new star was farther than the Moon, probably *much* farther, and this clearly falsified Aristotle's contention of the changelessness of the heavens.

Tycho considered himself to be a nobleman and had a very high and mighty opinion of his social status (even though he condescended to marry a lower-class woman and to have a very happy married life with her). Ordinarily, he would have considered it far beneath his dignity to write a book, but he was so overcome by the importance of the phenomenon of a new star and of the manner in which it disproved Aristotle that he wrote a book of fifty-two large pages that was published in 1573. It contained all the observations and measurements he had made of the star and all the conclusions he had come to. It instantly made him the most famous astronomer in Europe.

The book was written in Latin, the universal language of European scholars at the time, and it had a long title, after the fashion of the day. The book is usually referred to, however, by a short version of the title: *De Nova Stella* ("Concerning the New Star").

As a result of this title, the kind of new star I have been talking about in this essay is invariably known as a "nova," which is the Latin word for "new." The plural, in Latin, is "novae," but it is no longer fashionable to use Latin plurals, and one generally speaks of "novas" now.

Naturally, after Tycho's great success, other astronomers began to keep a sharp eye out for novas.

In 1596, for instance, the German astronomer David Fabricius (1564–1617), a friend of Tycho's, located a star in the constellation of Cetus that he had not noticed there before. It was a star of only the third magnitude, meaning that it was of only middling brightness (stars of the sixth magnitude are the dimmest that can be seen with the unaided eye), and Fabricius deserves credit for noticing it.

The fact that he hadn't seen the star before might not mean that it was really a nova, of course. It might have been there all the time and he might simply not have noticed it before. It wasn't on the star maps (Tycho had prepared the best one yet), but even Tycho's map wasn't perfect.

There was an easy solution, however; Fabricius had only to keep observing the star. He noted that from night to night the star dimmed, until it finally disappeared. That made it a nova all right, as far as Fabricius was concerned, and he announced it as such. It was so dim a nova, however, that it didn't cause much of a stir.

Another German astronomer alive at the time was Johannes Kepler (1571–1630). He had worked with Tycho in the last years of the older astronomer's life and Kepler was to prove to be an even more remarkable scientist for reasons outside the limits of this essay.

In 1604, Kepler noted a bright new star in the constellation Ophiuchus. It was considerably brighter than Fabricius's nova, for it was as bright as the planet Jupiter. That meant it wasn't as bright as Tycho's nova had been, but it was bright enough. Kepler observed it for as long as it remained visible, and it was a whole year before it faded to invisibility.

At this time, astronomy was on the verge of a remarkable revolution. The telescope was about to be invented and observations were soon to be made that wouldn't have been conceivable in previous ages. What's more, the telescope was to be the forerunner of other technological advances that were greatly to enhance the power of the astronomer to study the Universe, until finally we developed the huge radio telescopes and the interplanetary probes of today.

How much better we can study these novas in our time than Tycho and Kepler could in their time!

Nevertheless, it is the misfortune of astronomers that Kepler's nova of 1604 was the very last new star of planetary brightness to appear in the sky. Since then—nothing.

And yet, even so, knowledge concerning the novas continued to advance, as I shall explain in the next chapter.

12

Brightening Stars

I was asked recently to write an essay on the new motion picture *Star Trek IV: The Journey Home.* A young lady, working for the people who wanted the essay, undertook to obtain a pair of tickets to the preview for me and for herself. She would then tell me where and when to meet her.

The days passed and I heard nothing further about the project. Early on the very day of the preview she finally phoned me. It seemed she had had a great deal of trouble obtaining the tickets.

"Why?" I asked. "You represent an important outlet and the movie people should be overjoyed at the thought of having me write about the picture, since they must know I'm fond of *Star Trek.*"

"That's the point," said the exasperated young lady. "When they seemed reluctant to let me have tickets, I said to them, 'Don't you want Isaac Asimov to write about the picture?' And the girl at the other end said, 'Who is Isaac Asimov?' Can you *imagine?*"

I laughed and said, "Of course I can imagine. I estimate that about one American in a hundred has heard of me. You just encountered one of the other 99 percent. What did you do?"

She said, "I Xeroxed the several pages of *Books in Print* that listed your books, and messengered it to the girl with a note that said, '*This* is Isaac Asimov.' She phoned at once and said the tickets would be waiting at the box office."

The young lady and I went to the box office in due time, found the tickets indeed waiting for us, and went in to see the picture (which I thoroughly enjoyed). And when, as was inevitable, people near my seat began to pass me their programs for my signature, my companion fumed, "*How* could that person not know you?"

And I said, "Please. I welcome incidents like that. It helps keep my feet on the ground."

Nevertheless, I don't want it to happen *too* often, so I'll keep on writing these essays, and perhaps one or two more people will hear of me as a result.

In the previous chapter, I wrote about novas, or "new stars," that suddenly blazed out in the heavens. I ended with the nova of 1604, observed by Johannes Kepler, and mentioned that it was the very last nova to appear in the heavens with a brightness that rivaled that of planets such as Jupiter or Venus.

Now let us move on. In 1609, the Italian scientist Galileo Galilei (1564–1642) constructed a telescope after hearing rumors that such a device had been invented in The Netherlands. He then did something that the earlier telescopists did not think of doing. He turned it on the heavens.

One of the first things that Galileo did with his telescope was to look through it at the Milky Way. He found the Milky Way to be not just a luminous mist, but an aggregation of very faint stars, individually too dim to be seen with the unaided eye. In fact, wherever Galileo turned his telescope, he found that the instrument brightened all the stars and made visible numerous stars that were ordinarily too dim to see.

To us that does not seem startling. After all, there is a vast range of brightness in the heavenly bodies, from the Sun itself right down to the dimmest stars we can see. Why should not the range be extended to still fainter stars too dim to make out? To us, it might seem, in hindsight, that Galileo's discovery was merely a confirmation of something that was so obvious it scarcely needed confirmation.

That was not the way it seemed in Galileo's time, however. People were then quite certain that the Universe was created by God for the specific use of human beings. Everything in existence was designed to make human life possible, or to add to human comfort, or to serve to develop the human character and exercise the human soul, or, at the very least, to inculcate an improving moral lesson.

But what possible use could invisible stars have?

The impulse must have been to suppose that stars that were visible only in a telescope were artifactitious; that they were somehow created by the telescope, and were illusions that did not exist in reality. Indeed, there is a well-known story that when Galileo discovered the four large satellites of Jupiter, one scholar pointed out that since these satellites were nowhere mentioned in the writings of Aristotle, they did not exist.

However, the use of the telescope spread. Many were constructed, and the same stars seen and reported by Galileo were seen and reported by other astronomers as well. Eventually, it had to be accepted that God had created invisible stars and this was the very first hint that perhaps the Universe had not been created with human welfare as its primary object (a point I have never seen stressed in histories of science).

The discovery couldn't help but alter the way astronomers looked at novas. As long as only visible stars were thought to exist, a star that became visible where none had been visible before had to be thought of as coming into existence for the first time. It was a new star (and, as I pointed out last month, the very word "nova" is Latin for "new"). Again, when a nova faded to invisibility, it had to be thought of as passing out of existence.

If stars could exist, however, that were too dim to be seen without a telescope, it might well be that a nova was a star that was always in existence. It might have simply been too dim to see, but brightened until it became visible to the unaided eye and then, eventually, dimmed until it was too dim to be seen without a telescope.

A nova might not be a new star, then, but merely one whose brightness was not constant, as was the case for ordinary stars. A nova was a "variable" star.

This was soon shown to be so in connection with the apparent nova that had been sighted in 1596 by David Fabricius (as mentioned in the previous chapter) in the constellation of Cetus. At its peak, it was a star of only middling brightness—of the third magnitude—but after a while it disappeared. That made it a nova in those pretelescope days.

In 1638, however, a Dutch astronomer, Holwarda of Franeker (1618–51), sighted a star precisely where Fabricius had seen his nova forty-two years earlier. Holwarda watched it fade, apparently disappear, and then eventually return. But Holwarda had the advantage of a telescope and, as he watched the star, he found that it never really disappeared. It grew dimmer, yes, until it couldn't be seen by the eye alone, but it remained visible at all times when viewed through the telescope.

A variable star, in those days, was as revolutionary as a new star. The old Greek doctrine of the changeless perfection of the heavens was upset as completely by the one as by the other.

It turned out that, at its brightest, the star observed first by Fabricius and then by Holwarda, was about 250 times as bright as it was at its dimmest, and it oscillated between these extremes every eleven months or so. The German astronomer Johannes Hevelius (1611–87) gave this star the name of Mira (Latin for "wonderful") as a tribute to its astonishing property of variability.

Mira was the first variable star to be discovered, but as time went on, others were noted as well. Most variables turned out to be less variable than Mira, however.

In 1667, an Italian astronomer, Geminiano Montanari (1633–87), noticed that Algol, a star in the constellation Perseus, was variable. The variability was extremely regular, the star going through a cycle of brightening and dimming every 69 hours. At its brightest, Algol was only three times as bright as it was at its dimmest.

In 1784, an English astronomer, John Goodricke (1764–86), discovered that the star Delta Cephei, in the constellation Cepheus, varied with a regular cycle of 5.5 days, but was only twice as bright at its brightest as at its dimmest.

Many such variable stars are now known, and it might easily be argued that novas were variables, too. However, considering how brilliantly they shone, their brightness must alter to a much greater degree than was true of ordinary variables. Furthermore, since novas such as those observed by Tycho Brahe and Kepler seemed to appear once and were then forever invisible, they must be very *irregularly* variable.

All this indicated that there must be something very unusual about novas and it was a source of frustration among astronomers that, although they now had telescopes at the ready, no bright nova appeared in the sky after 1604.

In fact, even comparatively dim novas didn't appear (or, at least, weren't sighted) for a long time. In 1848, however, the English astronomer John Russell Hind (1823–95) observed a nova in the constellation of Ophiuchus. It didn't even reach fourth magnitude, however, so it was a rather dim star and would certainly have gone totally unnoticed in the days when fewer astronomers studied the sky, and did so in lesser detail.

Hind's nova was not an ordinary variable star because, having faded out, it didn't brighten again. There was no clear cycle of variability. It was a "one-shot," in other words, something that seemed at that time to be the key characteristic of a nova.

Three or four other such dim novas were detected in the remaining years of the nineteenth century. One of them was detected in 1891 by a Scottish clergyman and amateur astronomer, T. D. Anderson. It was only of the fifth magnitude.

Then, on the night of February 21, 1901, Anderson discovered a *second* nova, while walking home from some social engagement. This one, in the constellation of Perseus, came to be called "Nova Persei."

Anderson had caught it early and it was still brightening. Two days later, it reached its peak at a magnitude of 0.2, which was brighter than first magnitude. That made it as bright as Vega, the fourth brightest star. It was still well short of planetary brightness, but it was the brightest nova in three centuries.

By now, however, astronomers had the technique of photography at their disposal and this made it possible to find out something about novas that could not have been managed earlier.

The region of the sky in which Nova Persei shone had been frequently photographed, and, by looking at photographs taken before the nova had appeared, astronomers found that in the very spot in which Nova Persei was later shining there had existed a very dim star of the thirteenth magnitude. As they watched Nova Persei dim, they found that it eventually returned to the thirteenth magnitude.

In four days, it turned out, Nova Persei had increased its brightness 160,000 times, and, after a few months, all that added brightness was lost again. It was indeed an extreme variable, and its behavior was widely different from that of ordinary variables.

What's more, the camera could, by taking a long exposure, reveal detail that would escape the mere eye, even when reinforced by a telescope.

Some seven months after Nova Persei had blazed away in the sky, a long-exposure film of the dim star into which it had faded revealed a faint fog of light around it that gradually, over the weeks and months, grew in size. There was clearly a thin cloud of dust about the star that was reflecting light and expanding. By 1916, fifteen years later, the cloud had grown thicker and was continuing to expand outward from the star in all directions.

It seemed clear that the star had suffered a titanic explosion that had expelled gases. Nova Persei (and presumably other novas) were therefore said to belong to a group of stars called "eruptive variables," or "explosive variables." Such names, however, although descriptive and picturesque, could not and did not replace the older, shorter, and time-honored "nova."

A still brighter nova was seen by several different observers on June 8, 1918, in the constellation Aquila. At that time, it was a first-magnitude star, and two days later it was at its peak, shining with a magnitude of —1.1, or with almost the brightness of Sirius, the brightest star.

Nova Aquilae appeared during World War I, just as the Germans' last great offensive on the Western Front was beginning to run out of steam. Five months later, Germany surrendered and Nova Aquilae was called "the star of victory" by the Allied soldiers at the front.

Again, Nova Aquilae could be seen in photographs

taken before it had exploded. It was about three times as bright at its peak as Nova Persei had been (and no nova so bright has been seen since), but Nova Aquilae had been brighter to start with and brightened only 50,000 times as it exploded.

As it happened, Nova Aquilae had had its spectrum photographed before it had become a nova and, to this day, it is the *only* nova to have had its prenova spectrum recorded. The spectrum showed it to be a hot star with a surface temperature twice that of our Sun. This makes sense, for even without knowing anything about the details of a stellar explosion, it would seem logical that a hot star would be more likely to explode than a cooler one would.

In December 1934, a nova appeared in the constellation Hercules that managed to reach a magnitude of 1.4. Nova Herculis was not as bright as either Nova Persei or Nova Aquilae, and it would not have attracted much attention, except that after it returned to the thirteenth magnitude from which it had started four months earlier, it suddenly began to brighten again. After another four months, it was almost bright enough to be seen with the unaided eye. It was not till 1949 that it returned to the thirteenth magnitude a second time. Apparently, stars could brighten oftener than once and astronomers began to speak of "recurrent novas."

The most recent notable nova appeared in the constellation Cygnus on August 19, 1975. Nova Cygni brightened with unusual speed, becoming 30 million times brighter in the course of a single day, and reaching the second magnitude. It faded rapidly and was lost to the unaided sight within three weeks. Apparently, the faster and more extreme the brightening, the faster and more extreme the dimming.

And yet none of these novas I mentioned as having appeared during the days of the telescope are nearly as important as one I haven't mentioned; one that, at its peak, might have been just barely bright enough to be seen by the unaided eye.

This particular nova, which appeared in the constellation of Andromeda, may have been observed for the first time, on August 17, 1885, by a French astronomer, L. Gully. He was testing a new telescope, which turned out to be defective, and so he didn't feel he ought to make a fuss about sighting a new star that might not really be there.

An Irish amateur astronomer, I. W. Ward, may have observed the star on August 19, but again no fuss was made at the time, and he advanced his claim only later on.

The official discoverer was the German astronomer Ernst Hartwig (1851–1923), his first observation of the nova being on August 20, 1885. He judged it to be of the seventh magnitude, and possibly nearly the sixth magnitude.

The Moon, however, was almost full and observation was difficult. Hartwig decided to make further observations before announcing the new star but (of course) a week's worth of cloudy weather promptly intervened. Finally, on August 31, he sent off an official report. At once, other astronomers turned their telescopes on Andromeda.

The star at that time was still up in the region of the

seventh magnitude. Until that time no nova had been seen quite that dim and so there was no thought at first that that was what it was. It seemed an ordinary variable star. A variable star is named for its constellation and is given a letter prefix, starting with R and progressing up the alphabet. Since Hartwig's star was the second variable star to be recorded in Andromeda, it received the name S Andromedae.

By the end of August, however, the star was beginning to fade rapidly and it continued to fade until, half a year later, it had sunk to the fourteenth magnitude. It had been a nova, though an extraordinarily dim one, but it kept its name.

S Andromedae, however, was not merely in Andromeda, but in the center of an object within the constellation, an object called the "Andromeda Nebula"—and that is a story in itself.

The Andromeda Nebula can be seen with the unaided eye as a dim, somewhat fuzzy "star" of the fourth magnitude, and its position was noted by some of the Arabic astronomers of medieval times.

The first person to view it through a telescope, in 1611, was the German astronomer Simon Marius (1573–1624). It was then clearly seen to be no star. It was not a pointlike twinkle of light, but an extended misty object like a tiny cloud in the sky. (The word "nebula" is Latin for "cloud.")

The fuzzy objects that seemed most important to the astronomers of the eighteenth century were comets, but the Andromeda Nebula, and other objects like it, were *not* comets. A comet shifted position in the sky, changed its shape and brightness, and so on. The various nebulas, however, were changeless and motionless. Nevertheless, such nebulas were sometimes sighted by enthusiastic astronomers who thought they had discovered a new comet, and then found that they were wrong.

The most important comet-hunter of the eighteenth century was the French astronomer Charles Messier (1730–1817), and he resented being fooled in this way.

In 1781, therefore, he began to make a catalog of all the cloudy objects in the sky that might be mistaken for comets. His intention was to have every comet-hunter, before announcing a new comet, check his finding against the catalog and make sure he had not been fooled. Messier numbered the objects in the catalog (there were eventually 102 of them listed) and they are sometimes still known by that number with "M" (for Messier) as a prefix.

You can be sure that Messier included the Andromeda Nebula in his catalog. It was in thirty-first place, so that the Andromeda Nebula is often referred to as M31 in consequence.

The Andromeda Nebula puzzled astronomers. The most familiar fogginess in the sky was, of course, the Milky Way, and Galileo had shown that it was composed of very faint stars that, without the telescope, melted into a luminous haze.

In the Southern Hemisphere, one can see two cloudy patches that look like detached bits of the Milky Way. They were first sighted by Europeans in 1519, in the course of the expedition of Ferdinand Magellan (1480–1521), which followed the coast of South America to its southernmost reaches in the course of its trailblazing circumnavigation of the Earth. The patches are called the "Magellanic Clouds" in consequence, and these, too, could be seen by telescope to be made up of masses of faint stars.

The Andromeda Nebula, however, though it seemed to resemble the Milky Way and the Magellanic Clouds in appearance, could not be resolved into stars by any telescope of the eighteenth century (or of the nineteenth, for that matter). Why was that?

The first to express a useful idea on the subject was the German philosopher Immanuel Kant (1724–1804). In 1755, he reasoned that it must be that the Andromeda Nebula, and other similar patches of cosmic fog, were indeed composed of stars, but were so far away, so much farther away than either the Milky Way or the Magellanic Clouds, that even the best telescopes astonomers had were insufficient to separate the fog into stars. He spoke of the nebulas as "island universes."

Kant was right, dead right, in this, but it made no impression on the world of astronomy. It was too far ahead of its time. Astronomers in the eighteenth century had not yet determined the distance of any star, but there was a growing feeling that they must be very far away. The English astronomer Edmund Halley (1656–1742) had been the first to speak of stellar distances in terms of what we now call "light-years."

Astronomers, however, had been living in a small Universe all through history. The Universe had been visualized as barely large enough to hold what we now call the Solar System—and a Solar System that was viewed as much smaller than we now know it to be. To extend one's horizon to light-years was hard enough, but when Kant spoke of distances far greater still, so great that even telescopes couldn't make out individual stars, that was too much. Astronomers shuddered and turned away.

Less visionary, and, therefore, more acceptable, was a second view, that of the French astronomer Pierre-Simon de Laplace (1749–1827). He suggested, in 1798, that the Solar System was, to begin with, a vast spinning cloud of gas and dust that slowly condensed, with the center of the cloud becoming the Sun and the outskirts forming the planets. (Kant had actually made a similar suggestion in the same book in which he spoke of island universes, but Laplace went into greater detail.)

Laplace thought he could strengthen the argument by pointing out an example of a star and planetary system actually in formation, and the Andromeda Nebula seemed tailor-made for that. There was the explanation for its glow. A star was beginning to shine at its center and it illuminated the vast cloud of dust and gas that still surrounded and masked it. Telescopes could not resolve that cloud into separate stars because it didn't consist of separate stars. It was only one star, not yet fully formed.

Because of his use of the Andromeda Nebula as an example, Laplace's notion was called the "nebular hypothesis."

If Laplace were correct, then, the Andromeda Nebula was not an incredible distance away, as was demanded by Kant's notion, but had to be quite close to us, since so small an object as a single planetary system would not seem so large otherwise.

During the nineteenth century, the Andromeda Nebula grew steadily less unusual. As the skies were searched with better and better telescopes, it turned out that there were quite a number of nebulas that were luminous and that yet showed no signs of a star on even the closest examination.

The Irish astronomer William Parsons, 3rd Earl of Rosse (1800–67), paid particular attention to these nebulas and, in 1845, noted that a number of them seemed to have distinctly spiral structures, almost as though they were tiny whirlpools of light. The most spectacular example was one of the items on Messier's list—M51. It looked for all the world like a pinwheel and soon came to be known as the "Whirlpool Nebula." Astronomers began to speak of "spiral nebulas" as a not uncommon class of objects in the sky.

As the nineteenth century progressed, it began to be possible to photograph the nebulas with long exposures so that more detail could be seen than by eye alone.

In the 1880s, a Welsh amateur astronomer, Isaac Roberts (1829–1904), took a large number of such photographs. In 1888, he was able to show that the Andromeda Nebula had a spiral structure. This had not been noted before because the Andromeda Nebula was seen much more nearly edge-on than the Whirlpool Nebula.

Roberts pointed out that if successive photographs of nebulas over a period of years showed changes that indicated they were rotating at a measurable speed, then they would have to be close by. Anything as far away as Kant's island universes would take millions of years to show measurable changes. In 1899, Roberts claimed to have seen rotational changes in his many photographs of the Andromeda Nebula.

Also in 1899, the spectrum of the Andromeda Nebula was taken for the first time, and it proved to have all the characteristics of starlight, which might indicate there was a developing star within it.

Between the claims that the Andromeda Nebula was visibly rotating and the fact of its starlike spectrum, the matter seemed settled. In 1909, the English astronomer William Huggins (1824–1910) insisted there was no further doubt that the Andromeda Nebula was a planetary system in a late stage of development.

But one little point remained unsettled, and that was the matter of S Andromedae. This is a subject we will turn to in the next chapter.

13

Super-Exploding Stars

Last week, my dear wife, Janet, took me to an old mansion of colonial origin right here in Manhattan. I wouldn't have believed that any such relic remained on this island, but there it was. We paid a small sum (well worth it), signed the visitor's book, and were shown through by a very pleasant woman.

When we were almost done, another woman approached diffidently. She was carrying paperbacks of my first three Foundation novels. (I can recognize my own books in any edition as far as I can see them.)

"Dr. Asimov?" she said.

I said, "Yes, ma'am?"

She said, "My son is a great fan of yours and when I saw your name in the visitors' book, I called him up and said I thought you were in the house, but that I couldn't be sure which one you were. He said, 'Is there someone there with big white sideburns?' I said, 'Yes, there is.' He said, 'That's Dr. Asimov,' and he brought over these books."

So I signed them.

I always say that I have three trademarks: my bolo ties, my dark-rimmed glasses, and my white sideburns. However, anyone can wear bolo ties and dark-rimmed glasses. It's the white sideburns that really give me away because so few people care to sport such facial adornments. Fortunately, I am an unselfconscious, extroverted individual and I don't mind being recognized, so I don't intend to shave them off.

At the time of this incident, as it happens, I already knew I was going to be writing this essay, in order to conclude the subject I had been discussing in the previous two, and it occurred to me that I was going to be discussing a star that, in a sense, gave away the Andromeda Nebula, as my sideburns gave me away. Let me explain—

I pointed out in the previous chapter that, early in the twentieth century, there was a controversy concerning the Andromeda Nebula. There were those who thought it a huge and very distant collection of individually invisible stars, lying far outside our own Galaxy. If so, the Andromeda Nebula was certainly one of many such objects and the Universe was therefore far huger than astronomers, generally, realized it was as the twentieth century opened.

There were others who thought that our Galaxy (plus the Magellanic Clouds) did indeed make up essentially all of the Universe, and that the Andromeda Nebula, and all other such bodies, were relatively small, nearby clouds of dust and gas, existing inside our own Galaxy. Some even thought that such nebulas represented single planetary systems in the process of development.

In the argument between the "far-Andromedas" and the "near-Andromedas" (my own names for the two sides), the near-Andromedas seemed to have won the game hands down. The key piece of evidence was the photographs of the Andromeda Nebula, taken over the years, that seemed to show it to be rotating at a detectable speed. If it were far outside our own Galaxy, any motion would be immeasurably small, so that any detection of measurable rotation meant that it had to be a nearby object.

That left one undecided problem. As I explained in the previous essay, a star had appeared in August 1885 in the Andromeda Nebula, and was referred to as "S Andromedae." Since it appeared where no star had been detected before, and since it had gotten too dim to be seen after seven months, it was a nova. It was, however, the dimmest nova that had ever been detected, for even at its brightest, it had reached only the very borderline of naked-eye visibility. It very likely would never have been detected at all had it not appeared right in the middle of the blank fog of the Andromeda Nebula.

No one paid very much attention to it at the time, but as the controversy over the Andromeda Nebula grew hot, S Andromedae took center stage. If the nova were actually located in the Nebula, then the Nebula was not likely to be a mere cloud of dust and gas. It was more likely to be a cluster of very dim stars, in which one star exploded and grew bright enough to make out with a telescope. That would be a strong point in favor of the far-Andromeda view.

The catch to this argument was that there was no way of showing that S Andromedae was actually part of the Andromeda Nebula. It might merely be a star that existed in the *direction* of the Nebula, but was much closer to us than the Nebula was. Since we don't see the sky three dimensionally, a nearby S Andromedae that happened to be in the direction of the Andromeda Nebula would seem to our eyes to be part of the Nebula even though it wasn't.

But if the Andromeda Nebula were comparatively close and S Andromedae were closer still, why should it have been so dim?

Well, why not? There are lots of nearby stars that are dim. Barnard's Star is only six light-years away (only the Alpha Centauri system is nearer) and yet Barnard's Star can only be seen by telescope. One of the stars of the Alpha Centauri system itself, Alpha Centauri C, or "Proxima Centauri" (discovered in 1913), is the nearest of all known stars, yet is far too dim to be seen by the unaided eye.

There are many very dim stars and S Andromedae might well be one of them and might be none too bright even when it became a nova—so the near-Andromedas still had the best of it.

But then came 1901 when, as I mentioned in the previous chapter, Nova Persei flashed out, the brightest nova in three centuries. As it happened, the telescope showed a cloud of gas and dust around it (the result of the star's explosion) and a circle of illumination seemed to be spreading outward with time. Astronomers felt it was light traveling outward from the star and illuminating the dust farther and farther outward. The actual speed of light was well known, and from the apparent speed with which light was expanding outward, it was easy to estimate the distance of the nova. Nova Persei turned out to be about 100 light-years away.

That's not very far; only about 25 times as far as the nearest star. No wonder Nova Persei appeared so bright.

What, then, if all novas, in exploding, ended up having about the same luminosity? They might all be of equal apparent brightness if all were at equal distance but, since they were probably at wildly unequal distances, those were brighter which were closer.

In that case, if S Andromedae reached the same luminosity at its peak that Nova Persei did, and was as dim as it seemed only because of its greater distance, that distance could be calculated. Then, if S Andromedae was not actually part of the Andromeda Nebula, that meant the Nebula must be farther away still, perhaps much farther away.

The far-Andromeda view brightened slightly, but not very much. After all, this argument rested on a very shaky underpinning. What right had anyone to make the assumption that all novas reached about the same luminosity? There was no compelling reason to suppose this. It was just as reasonable to suppose that faint stars gave rise to faint novas and that S Andromedae was a very faint star. It might be closer than Nova Persei and yet be much fainter in its nova stage.

The near-Andromeda side still seemed to have the better of it.

* * *

One American astronomer was stubbornly far-Andromeda in his belief and he refused to accept this last argument.

He was Heber Doust Curtis (1872–1942). He began his academic life by studying languages and becoming a teacher of Latin and Greek. The college he taught at had a telescope, however, and Curtis grew interested in it, and then in astronomy, which he had never studied in school. In 1898, he switched careers and became an astronomer, getting his Ph.D. in the subject in 1902.

In 1910, he was put to work on nebular photography and, naturally, he was drawn into the controversy over whether the nebulas were distant objects beyond the Galaxy or were nearby objects.

One of the arguments in favor of thinking the nebulas to be part of our Galaxy was this: If they were outside our Galaxy, then they should be scattered all over the sky indiscriminately, since there seemed no reason why they should be in one direction rather than in another. In actual fact, though, the nebulas were found in greater and greater numbers the farther one explored away from the line of the Milky Way. This, it was argued, showed that the nebulae were likely to be part of the Galaxy, since objects within our Galaxy might not form near the Milky Way for some reason or other, whereas objects outside our Galaxy should have no reason at all for being influenced, one way or the other, by some feature within our Galaxy.

Curtis, however, in photographing the various nebulae, noticed that many of them possessed dark, opaque clouds located on the outskirts of their often flattened-pancake masses.

It seemed to Curtis that the outer rim of our own Galaxy (marked off by the Milky Way) might also have dark, opaque clouds, and, indeed, a number of them could be seen in the Milky Way. Curtis argued, therefore, that the nebulas were indeed distributed evenly over the sky, but that the dark clouds in the neighborhood of the Milky Way hid many of them and made it *seem* there were more of them far from the Milky Way than near it.

And if that were so, then this particular argument for having the nebulas part of our Galaxy bit the dust, and Curtis's far-Andromeda views were strengthened.

He next began to reason thus— The Andromeda Nebula was the largest of the nebulas and the brightest (next to the Magellanic Clouds, which existed just outside our Galaxy and were its satellites, so to speak). Except for the Magellanic Clouds, the Andromeda Nebula was the only one visible to the unaided eye. This probably meant that it was the closest nebula outside the Magellanic Clouds and was the most likely to give observing astronomers important detail.

If the Andromeda Nebula, then, was a very distant collection of stars, so distant that the component stars could not be individually seen, those stars would be more nearly individually visible than the stars of any other nebula. It would follow that if one of the stars of the Andromeda Nebula brightened as a nova would, it might become visible, and that this would account for S Andromedae. This might not be true of farther nebulas where individual stars would be entirely too faint for even novas to become visible.

Beginning in 1917, then, Curtis began a careful and persistent series of observations of the Andromeda Nebula to see if he could find other novas—and he did. He found that stars appeared and then disappeared, dozens of them. That they were novas was unquestionable, but they were amazingly dim. They could just barely be made out with his telescope. This was to be expected if the Andromeda Nebula was truly far away.

Could it be, though, that Curtis was merely seeing very faint novas in the direction of the Andromeda Nebula and that none of them were actually *in* the Nebula. If that were so, the Nebula might still be merely a cloud of dust and gas.

To Curtis, however, that seemed quite out of the question. Nowhere else in the sky could one find such a crowd of very faint novas in a small area equivalent to that covered by the Andromeda Nebula. As a matter of fact, there were more novas seen in the direction of the Nebula than in all the rest of the sky put together. There was simply no reason why this should be if the Andromeda was merely an unremarkable cloud of dust and gas.

The only logical explanation was that the novas were in the Andromeda Nebula and that their great number was merely a reflection of the far vaster number of stars, generally, that existed there. In other words, the Andromeda Nebula was a galaxy like our own and, in that case, it must lie very far away. Its great distance, then, would account for the extraordinary faintness of the novas.

Curtis became the outstanding astronomical spokesman for the idea of the far-Andromeda.

But what about the key observation that supported

the near-Andromeda idea; the fact that the Andromeda Nebula was observed to rotate. That rested on nineteenth-century observations that might be questionable, but in the early twentieth century, the observations were strengthened.

At just about the time that Curtis was detecting novas in the Andromeda Nebula, a Dutch-American astronomer, Adriaan van Maanen (1884–1946), was carefully observing nebulas and checking their apparent rotation. He was working with better instruments and making better observations than his predecessors had, and he reported that he had definitely detected a measurable rate of rotation in the Andromeda Nebula, and in several other nebulas as well.

What it amounted to was this— If Curtis had truly detected faint novas in the Andromeda Nebula, then it was simply impossible that van Maanen had actually detected a very tiny speed of rotation for the Nebula. And if van Maanen had actually detected rotation of the Nebula, then it was simply impossible that Curtis had detected numerous faint novas in it. The two observations were mutually exclusive and which should one believe?

No clear decision could be made. Both Curtis and van Maanen were observing something that was just at the very limits of observation. In either case, a very slight error in the instrument or in the observer's judgment might wipe out the observation. This was all the more true since both astronomers were detecting something they very much wanted to detect and were sure they would detect. Even the most honest and scrupulous scientist could be swayed into observing something that wasn't there to observe if he were emotionally motivated to observe it. So although only one of the two could be correct, there seemed to be no way of deciding which one it might be.

One of the most prominent American astronomers of the time was Harlow Shapley (1885–1972). It was Shapley who had worked out the true vastness of our own Galaxy (indeed, he had overestimated its size somewhat) and had shown that our Sun was not at its center but was located in the outskirts.

Perhaps, as the enlarger of the Galaxy, Shapley didn't quite like the notion of finding the Universe containing a great many galaxies, thus reducing our own to insignificance again. It is difficult, however, to argue psychological motivations, and probably unfair. Shapley also had objective reasons to favor the near-Andromeda idea.

Shapley was a very close and longtime friend of van Maanen, and an admirer of his astronomical work. It was only natural, then, for Shapley to accept van Maanen's observations of the rotation of the Andromeda Nebula. So did most of the astronomical community, and Curtis found himself in a minority.

On April 26, 1920, Curtis and Shapley held a well-publicized debate on the matter before a crowded hall at the National Academy of Sciences. Since Shapley was far better known than Curtis was, the astronomers in the audience expected the former to have no trouble in establishing his point of view.

Curtis, however, was an unexpectedly effective speaker, and his novas, in their dimness and their number, proved a surprisingly powerful argument.

Objectively, the debate should be considered to have been a standoff, but the fact that Curtis had not been demolished, but had actually held the champion to a draw, was an astonishing moral victory. As a result, there developed a steadily growing opinion that Curtis had won the debate.

He did, in fact, win over a number of astronomers to the far-Andromeda viewpoint, but scientific issues are not settled by victory in debate. Neither Curtis's nor van Maanen's observations were sufficiently compelling to end the controversy. Something else was needed; new and better evidence.

The man who supplied it was the American astronomer Edwin Powell Hubble (1889–1953). He had at his disposal a new giant telescope with a mirror 100 inches in diameter—the most farseeing anywhere in the world up to that time. It was put into use in 1919 and, in 1922, Hubble began to use it to make time-exposure photographs of the Andromeda Nebula.

On October 5, 1923, he found, on one of these photographs, a star in the outskirts of the Andromeda Nebula. It was not a nova. He followed it from day to day and it turned out to be the kind of star known as a "Cepheid variable." By the end of 1924, Hubble had found thirty-six very faint variable stars in the Nebula, twelve of them Cepheids. He also discovered sixty-three novas, much like those that Curtis had earlier detected, except that Hubble, with the new telescope, could see them more clearly and unmistakably.

Hubble reasoned, much as Curtis had done, that all these stars found in the direction of the Andromeda Nebula could not exist in the space between it and us. They had to exist within the Nebula, which therefore had to be a conglomeration of stars.

In fact, Hubble's discoveries went beyond Curtis's

in a crucial way. Cepheid variables can be used to determine distances (a technique that Shapley had used very effectively in measuring the dimensions of our own Galaxy). And now Hubble used that same technique to demolish Shapley's stand on the matter of the Andromeda Nebula, because the Cepheids he had detected made the Andromeda Nebula seem to be about 750,000 light-years away. (As a matter of fact, in 1942, the German-American astronomer Walter Baade [1893–1960] refined the technique of measurement by Cepheids and showed that the correct distance of the Andromeda Nebula was about 2.3 million light-years.)

With that, the victory of the far-Andromeda view was complete. Van Maanen's observations had been wrong for some reason (perhaps instrument flaw) and no one has observed any measurable rotation in the Andromeda Nebula since. Indeed, from Hubble's time on, the structure has been named the Andromeda Galaxy, and the other "extragalactic nebulae" have also come to be called galaxies.

But there remained a problem. It had been S Andromedae, you will recall, that had been the nagging question that had kept astronomers wondering about the Andromeda Nebula. That nova had cast doubt on the Nebula being a nearby object.

However, now that that matter was settled and astronomers spoke of the Andromeda Galaxy, S Andromedae became a puzzle in the other direction. Earlier, astronomers had wondered about its dimness; now they wondered about its brightness. The more than one hundred novas observed in the Andromeda Galaxy were all extremely dim. S Andromedae was millions of times brighter than they were—all but bright enough to be made out by the unaided eye. Why was that?

Again, there were two possibilities. One was that perhaps S Andromedae had indeed flared up in the Andromeda Galaxy, but it just happened to be a few million times more luminous than ordinary novas. That seemed so unreasonable that almost no astronomer would believe it. (However, Hubble did, and, at the moment, his prestige was sky-high.)

The second possibility seemed more likely—that S Andromedae was not part of the Andromeda Galaxy, but, by a not-impossible coincidence, lay in the same direction as that body. If it were only a thousandth as far as the Andromeda Galaxy, it would naturally seem millions of times as bright as the dim, dim novas that *were* part of that galaxy. Most astronomers took this view.

You can't, however, settle a dispute of this kind by majority vote. Once again, there had to be new and better evidence, one way or the other.

A Swiss astronomer, Fritz Zwicky (1898–1974), pondered the problem. Suppose S Andromedae were part of the Andromeda Galaxy and had blazed up with a fierce light a few million times brighter than any ordinary nova would exhibit. Suppose that, in other words, S Andromedae was not merely an exploding star, but a super-exploding star, or a "supernova" (to use a term that Zwicky himself introduced).

If so, there had been one supernova noted in the Andromeda Galaxy and *many* ordinary novas. That made sense since anything that is an extreme in the direction of hugeness is bound to be far less numerous than things that are comparatively ordinary.

There was, therefore, little use in watching the Andromeda Galaxy, or any one galaxy, for another supernova. It could take decades, or centuries, to spot one in that way.

However, there were millions of distant galaxies so far away that ordinary novas could not possibly be detected under any conditions. Supernovas, on the other hand, would be seen in them. S Andromedae had shone with an intensity that was a respectably large fraction of all the light in the rest of the Andromeda Galaxy (provided S Andromedae had really been a part of that galaxy). If other supernovas were like S Andromedae, they, too, would shine with the concentrated light of an entire galaxy, so that no matter how far off a galaxy might be, as long as it was close enough to be seen at all, a supernova within it would also be seen.

Any one particular galaxy might have a supernova only at rare intervals, but every year, there might be supernovas showing up in one galaxy or another. An astronomer must therefore watch as many galaxies as possible and wait till he sees one of them (*any* one of them) which has grown a star as bright as itself that wasn't there before.

In 1934, Zwicky began a systematic search for supernovas. He focused on a large cluster of galaxies in the constellation of Virgo and watched them all. By 1938, he had located no fewer than twelve supernovas, one in each of twelve different galaxies of the cluster. Each supernova, at its peak, was almost as bright as the galaxy of which it was part, and each one of them had to be shining (at its peak) with a luminosity that was billions of times that of our Sun.

Could this observation be deceptive? Could Zwicky

have just happened to spot twelve ordinary novas that were much closer than the galaxies in which they seemed to exist, but that merely happened to be in the same direction as those galaxies?

No, that couldn't be. The twelve galaxies were very tiny patches in the sky and to have twelve novas, each of which was located in precisely the same direction as one of those galaxies, would ask far too much of coincidence. It was much more sensible to accept the notion of supernovas. Besides, additional supernovas were discovered in succeeding years by Zwicky and others. By now, over 400 supernovas have been detected in various galaxies.

Is it possible, then, that some of the novas seen in our own Galaxy have been supernovas?

Yes, indeed. It is not likely that an ordinary nova would be so close to us as to shine with a light exceeding that of the planets. A supernova could do so easily, however, even if it was quite far away.

Thus, the really bright novas I described in chapter 11 must have been supernovas. That includes the nova of 1054, Tycho's nova of 1572, and Kepler's nova of 1604.

The 1604 supernova was the most recent to have been visible in our own Galaxy. Since the development of the optical telescope, the spectroscope, the camera, the radio telescope, and rockets, there have been *no* supernovas in our own Galaxy that we could see. (There may have been some on the other side of the Galaxy, where they would be hidden by the opaque obscuring clouds between us and the galactic center.)

In fact, since 1604, the closest supernova we have

experienced was S Andromedae, and that was a century ago and was 2.3 million light-years away.

While no sane person would wish a supernova to erupt too near the Earth, we would be safe enough if one erupted, say, 2,000 light-years away. In that case astronomers would have a chance to study a supernova explosion in enormous detail, something they would dearly love to do.

Astronomers are, therefore, waiting for such an event, but that's all they can do—wait. —And gnash their teeth, I suppose.

[NOTE: *Less than a month after this essay was written, a supernova appeared, not in our own Galaxy, but in our very nearest neighbor, the large Magellanic Cloud. Astronomers were beside themselves with joy at having a supernova that was only 150,000 light-years away.]*

14

The Dead-End Middle

Last night I sat at the piano in the evening and tapped out tunes with one hand. I didn't have a piano available to me until the 1950s, but even at that late date I remembered perfectly well what I had been taught, in the fourth grade, about the staff, and the notes, and the sharps and flats. Beginning with that, once I had a piano, I tapped out the notes of familiar tunes (I have a good ear), and compared them with the musical notation. In this way, I gradually taught myself to read music—in a very primitive way.

So last night, as I listened to myself tap out "My Old Kentucky Home" and "The Old Folks at Home" and a few other simple ballads, *without* the music in front of me, I sighed and said to my dear wife, Janet, "If only I had had a piano available to me as a child, when I had time to fool around with it. I would probably have banged away at it till I could play chords and produce reasonable music by ear. Someone would surely have helped me over the rough spots and by the time I was an adult I would be playing well enough to amuse myself, even if not very well in an absolute sense."

Janet (who had had piano instruction as a child and who can play well enough to amuse herself) sympathized warmly, as she always does.

But then I searched for the bright side, since I hate feeling sorry for myself, and said, "Of course, it would have meant I would have wasted a great deal of time, and would have ruined a sizable segment of my life."

This Janet quite understood because she has long since learned that I consider any time not spent in writing as a waste (always barring the time I spend with her—if not excessive).

So I am making up for the time wasted at the piano last night by writing about that moment—and now, refraining from wasting any further time, I will continue to write, even if about something else.

—

We all know that we can get energy out of atomic nuclei if we tear them apart into smaller bits (nuclear fission), or if we squash them together into larger bits (nuclear fusion).

It might occur to someone, then, that it should be possible to get infinite amounts of energy by alternately tearing nuclei apart and then squashing them together again, over and over. Unfortunately, a malevolent nature has anticipated this plan and passed laws of thermodynamics against it.

Massive nuclei can indeed be split to produce energy but the fission products cannot be fused back to the original nuclei without reinserting at least as much energy as had been produced in the fission.

Again, light nuclei can indeed be fused to produce energy but the fusion products cannot be fissioned back to the original nuclei without reinserting at least as much energy as had been produced in the fusion.

If, then, we consider the spontaneous changes in the Universe, there is a tendency for massive nuclei to undergo fission and for light nuclei to undergo fusion. In each case, the change is one way.

The massive nuclei give off energy as they gradually become less massive; the light nuclei give off energy as they gradually become more massive. In either case, nuclei are produced that have a smaller energy content than the originals; and in either case it means that the particles making up the product nuclei are less massive, on the average, than those making up the original nuclei.

If we imagine this proceeding from the massive nuclei to the less massive, and from the light nuclei to the more massive, we can see that we must come across a nucleus, somewhere in between, which has a minimum energy content, and a minimum average particle mass. Such an in-between nucleus cannot give off any further energy by becoming either smaller or larger. It can undergo no further spontaneous nuclear change.

This dead-end middle is represented by the nucleus of iron-56, which is made up of 26 protons and 30 neutrons. This is the nucleus toward which all nuclear change is tending.

Let's try some figures—

The single particle of the nucleus of hydrogen-1 has a mass of 1.00797. The twelve particles of the nucleus of carbon-12 have an average mass of 1.00000 (it is this average that defines the nuclear unit of mass). The sixteen particles of the nucleus of oxygen-16 have an average mass of 0.99968. And the fifty-six particles of the nucleus of iron-56 have an average mass of 0.99884. (These are small differences of mass, but even a tiny loss of mass is equivalent to a comparatively huge gain of energy.)

Working from the other end, the 238 particles in the nucleus of uranium-238 have an average mass of 1.00021. The 197 particles in the nucleus of gold-197 have an average mass of 0.99983. The 107 particles in the nucleus of silver-107 have an average mass of 0.99910. You see, then, that from both directions, the nuclei are working their way down to iron-56 as the least massive per nuclear particle, and, therefore, possessing the least energy content, and being the stablest.

In our Universe, the predominant nuclear changes are fusion in character. After the first moments of the big bang, the Universe consisted of hydrogen plus helium (with very small nuclei) and nothing more. The entire history of the Universe in all the fifteen billion years since the big bang has consisted of the fusion of these small nuclei to larger ones.

In the process a substantial quantity of more massive nuclei has been formed, some in greater quantity than others (depending on the rates of various fusion reactions), including a quantity of iron that is considerably greater than that of other elements of similar nuclear mass. Thus, the core of the Earth is thought to be very largely iron; and this may also be true of the cores of

Venus and Mercury. Many meteorites are some 90 percent iron.—All this is because iron is the dead-end middle.

To be sure, nuclei of elements more massive than that of iron have also been formed, for they exist. There are conditions, you see, in which the nuclear fusions from hydrogen to iron proceed at such an enormously explosive rate that some of the energy doesn't have time to escape and is, instead, absorbed by the iron atoms, which are, in this way, kicked up the energy scale, so to speak, to nuclei as massive as uranium, and even beyond.

These heavier nuclei are present in only trace quantities in the Universe as a whole. In fact, in all the fifteen billion years of the Universe's lifetime, only a very small fraction of the original matter of the Universe has fused even into nuclei of iron and less. Of the nuclei making up the Universe, 90 percent is still hydrogen and 9 percent is still helium. Everything else, formed by fusion, makes up 1 percent, or less, of the whole.

Why is that? It is because fusion processes do not take place easily. In order for two nuclei to fuse they must collide with considerable force—but nuclei are protected by layers of electrons under ordinary conditions. Even if the electrons are removed, the bare nuclei are all positively charged and tend to repel each other.

In order for the fusion process to take place, therefore, a mass of hydrogen must be under great pressure and high temperature, conditions that are sufficiently extreme only in such places as the core of stars.

Enormous energies must be pumped into hydrogen

atoms to get rid of the electrons and then to smash the bare nuclei (individual protons) against each other despite the repelling force of their like charges. How, then, can we speak of fusion as a "spontaneous change," when it requires so much energy to make it happen?

That is because this energy is an "energy of activation," something that serves to begin the process. Once the process of fusion is begun, energy is liberated in quantities sufficient to keep it going, even though most of it is radiated outward. Fusion thus produces *much* more energy than the small amount required to start it, so that, on the whole, fusion is a spontaneous energy-producing reaction.

If this sounds too confusing, consider a friction match. Left to itself at room temperature, it would never give off energy. Strike it on a rough surface, however, and the heat of friction will raise its temperature to the point where the chemical head of the match will start to burn. The heat of the fire will then raise the temperature of the surrounding materials to the point where they will burn. This can continue indefinitely, so that a match, once lit, can begin a forest fire that will level countless acres.

Even at the center of a star, the fusion process goes on comparatively smoothly and slowly. Our Sun has been fusing at its core for nearly five billion years with very little external change, and will continue doing so for at least five billion additional years.

While our Sun fuses hydrogen to helium, it is said to be on the "main sequence." This lasts a long time because the hydrogen-to-helium fusion produces a vast amount of energy.

During all the billions of years on the main sequence, more and more helium accumulates at the Sun's core, which thus grows very slowly more massive. The accumulating gravitational field of the core grows more intense and compresses it more and more, raising its temperature and its pressure, until finally, these quantities have grown high enough to supply the necessary energy of activation to bring about the fusion of helium nuclei to still more massive nuclei.

Once helium fusion begins, what is left of the fusion process is comparatively short, because all the fusion processes beyond helium produce only about one fifth of the energy that the initial hydrogen-to-helium fusion produced. What's more, with helium fusion, the star begins to change its appearance drastically and is said to have "left the main sequence." For a variety of reasons, it expands mightily and, with the expansion, its surface (but not its core) cools and reddens. The star becomes a "red giant," and its life thereafter, as an object undergoing fusion, is short.

A star that has roughly the mass of our Sun will have its fusion processes brought to the pitch where its core consists chiefly of such nuclei as those of carbon, oxygen, and neon. To cause these to undergo further fusion, a temperature and pressure must be reached which the gravitational intensity of the star and its core cannot produce.

The star therefore cannot produce enough fusion energy, at this point, to keep it expanded against the remorseless inward pull of its own gravity, so it begins to contract. The contraction raises the pressure and temperature in the outer regions of the star, which are still composed largely of hydrogen and helium. These regions undergo rapid fusion and are blown away in a cloud of incandescent vapor. Most of the star collapses, however, and becomes a white dwarf made up almost entirely of carbon, oxygen, and neon—no hydrogen and helium.

White dwarfs are stable objects. They are not undergoing fusion; they just slowly leak away the energy they have, so that, very slowly, they cool and dim until they eventually don't radiate visible light at all, and become "black dwarfs." This process is so slow that it may be that in all the history of the Universe, no white dwarf has had a chance yet to cool all the way to a black dwarf.

But what if a star is considerably larger than our Sun; three or four, or even twenty or thirty times as massive? The more massive a star, the more intense its gravitational field and the more tightly it can compress its core. The temperature and pressure of the core can go far higher than they ever will in our Sun. Carbon, oxygen, and neon can fuse to silicon, sulfur, argon, and all the way to iron.

But at iron the process comes to a dead halt, for iron can spontaneously undergo neither fusion nor fission. The core's production of energy fades and the star begins to collapse. The collapse is much faster under the gravitational pull of a giant star than under that of an ordinary one, and the quantity of hydrogen and helium still existing is much greater in the giant. There is an explosion of much of the hydrogen and helium in a comparatively short time, and for a few days or weeks, the star shines with a luminosity about a billion times that of an ordinary star.

We call the result a "supernova."

The vast explosion of a supernova sends nuclei of all sizes into the interstellar spaces. Some of the nuclei are even more massive than iron, for enough energy is produced to kick an occasional iron nucleus uphill.

A supernova spreads quantities of massive nuclei through the interstellar clouds, which, to begin with, consist of hydrogen and helium only. A star formed out of clouds containing such massive nuclei (our own Sun, for instance) incorporates them into its own structure. The massive nuclei also find their way into planets of such stars, and into life-forms that develop on those planets.

However, the core of the exploding supernova, which contains most of the iron and other massive nuclei shrinks to a tiny neutron star, or to a still tinier black hole. The major portion of the massive nuclei thus remains in place and never escapes into interstellar space. We might wonder, then, if such supernovas can account for the quantity of massive nuclei we find in the Universe generally.

The kind of supernova I have described, however, is not the only kind.

For the last half century, about 400 supernovas have been studied. (All of these have been in other galaxies, for no supernova has been spotted in our own Galaxy since 1604, to the chagrin of astronomers.) These supernovas can be divided into two classes, which are called Type I and Type II.

Type I tends to be more luminous than Type II. Whereas a Type II supernova can reach a luminosity about a billion times that of our Sun, a Type I supernova can be up to 2.5 billion times as luminous as our Sun.

If this were the only difference, we would assume that particularly large stars exploded to form a Type I supernova, whereas somewhat smaller stars exploded to form a Type II supernova. This seems so obvious that it would be very tempting to look no further.

There are other differences, however, that upset this conclusion.

For instance, the dimmer Type II supernovas are found almost always in the arms of spiral galaxies. It is precisely in those arms that one finds large concentrates of gas and dust and it is in those arms that one therefore finds large and massive stars.

The brighter Type I supernovas, however, though sometimes found in the arms of spiral galaxies, may also be found in the central regions of those galaxies, as well as in elliptical galaxies, where there is little dust and gas. In such gas- and dust-free regions stars of only moderate size generally form. From their locations, then, it would seem that it is the Type II supernova that forms from the explosion of giant stars, while the Type I supernova forms from the explosion of smaller stars.

Again, a third difference is that the Type I supernovas, having passed their peak, grow dimmer in a very regular fashion, while the Type II supernovas dim irregularly. Here, too, we would expect a smaller star to behave more decorously than a larger star. The more gigantic explosion of a larger star would be expected to have a more chaotic history, with subexplosions and so on.

From both the matter of location and the manner of dimming, we would expect Type I supernovas to originate from smaller stars than is true of Type II supernovas. But in that case, why are Type I supernovas up to 2.5 times as luminous as Type II supernovas?

Another point. Smaller stars are invariably more common than larger stars. One would therefore expect that Type I supernovas, if they originate in smaller stars, would be more common than Type II supernovas; perhaps ten times as common. But that is not so! The two types of supernovas are about equally common.

A possible resolution of this problem arises from the spectra of these two supernova types, which yield results that are widely different. Type II supernovas have spectra that have pronounced hydrogen lines. That is to be expected of a giant star. Even if its core is choking with iron, its outer regions are rich in hydrogen, the fusion of which supplies the energy that keeps such a supernova blazing with light.

The Type I supernova, however, yields a spectrum that shows no hydrogen. Only such elements as carbon, oxygen, and neon show up. —But that is the makeup of white dwarfs!

Can a Type I supernova be an exploding white dwarf? In that case, why are there so few Type I supernovas? Can it be that only a minority of the white dwarfs explode so that Type I supernovas end up being no more numerous than Type II supernovas? Why should only a minority of them explode? And why should they explode at all? Haven't I said earlier in the essay that white dwarfs are very stable and slowly dim over the course of many billions of years, suffering no other change?

The solution to such questions arose out of a consideration of novas. (Not supernovas, just ordinary novas, which flare up to a luminosity of only 100,000 to 150,000 times that of the Sun.)

Such novas are much more common than supernovas, and they can't represent major explosions of a star. If they were, they would be red giants before the explosion, would be much brighter at peak explosion, and would fade away just about altogether afterward. Instead, novas seem to be ordinary main-sequence stars both before and after their moderate brightening, with little if any apparent change as a result of their adventure. Indeed, a particular star can be a nova over and over again.

But then, in 1954, an American astronomer, Merle F. Walker, noted that a certain star, eventually called DQ Herculis, which had gone through a nova stage in 1934, was actually a very close binary star. It consisted of two stars so close together that they were nearly touching.

Every effort was made to study each star of the pair separately. The brighter of the two was a main-sequence star, but the dimmer was a white dwarf! By the time this was determined, a number of other stars known to have gone nova at some time in their history were found to be close binaries, and in each case, it turned out that one of the pair of stars was a white dwarf.

Astronomers quickly made up their minds that it was the white dwarf of the pair that went through the nova change. The main-sequence star was the one ordinarily observed, and it went through no significant change, which was why the nova seemed to be the same before and after its brightening. The white dwarf of the pair was *not* ordinarily observed, so the whole significance of the nova was lost.

But not anymore. Here is what astronomers grew quickly convinced must happen—

We begin with two main-sequence stars that make up a close binary pair. The more massive a star, the more quickly it uses up the hydrogen at its core, so the more massive of the pair is therefore the first to begin to expand into a red giant. Some of its expanding material leaks over to the less massive partner, which is still on the main sequence, and its life is shortened in consequence. Eventually, the red giant collapses into a white dwarf.

Sometime afterward, the remaining main-sequence star, its life shortened, begins to swell into a red giant and becomes large enough for some of its mass to leak across into the neighborhood of the white dwarf. It spirals into an orbit (or "accretion disk") about the white dwarf. When enough gas has crowded into the accretion disk, the disk collapses and pours itself onto the surface of the white dwarf.

Mass falling onto the surface of a white dwarf behaves differently from that falling onto the surface of an ordinary star. The white dwarf's gravitational intensity at its surface is thousands of times greater than the gravitational intensity at the surface of a normal star. Whereas matter picked up by a normal star is merely added quietly to the star's mass, matter picked up by a white dwarf is compressed under the intensity of its surface gravity and undergoes fusion.

When the accretion disk collapses, there is, therefore, a sudden surge of light and energy and the binary system brightens a hundred thousand times or so. Naturally, this can happen over and over again, and each time it happens, the white dwarf becomes a nova and also gains mass.

However, a white dwarf can only have a mass equal to 1.44 times the mass of the Sun. This was shown by the India-born astronomer Subrahmanyan Chandrasekhar, in 1931, and that mass is called the "Chandrasekhar limit." (Chandrasekhar got a too-long-delayed Nobel prize in physics for this in 1983.)

A white dwarf is prevented from shrinking further by the resistance of electrons to further contraction. However, when the white dwarf passes the Chandrasekhar limit, the gravitational intensity becomes so great that the electron resistance is shattered and a new contraction does begin.

The white dwarf shrinks with catastrophic speed, and as it does so, all the carbon, oxygen, and neon nuclei that make it up fuse, and the energy this produces tears the star *completely* apart, leaving only gaseous, dusty debris behind. It is for that reason that a Type I supernova, originating in a less massive star, is more luminous than a Type II supernova, originating in a much more massive star. The white dwarf explosion is total and not partial, and the explosion is much faster than that of a giant star.

And the reason the Type I supernova is not more common is because not every white dwarf explodes. Those white dwarfs which are single stars or are far from their companion stars (as the white dwarf Sirius B is far from its companion, the main-sequence Sirius A) have little or no chance of gaining mass. It is only those white dwarfs that are members of close binaries that can gain enough mass to surpass Chandrasekhar's limit.

In this way, many of the differences in the characteristics of the two types of supernovas are explained—but one difference still puzzled. Why do the Type I supernovas dim in such a regular fashion, whereas the Type II supernovas dim irregularly?

In June 1983, a Type I supernova erupted in the relatively nearby galaxy M83. It was particularly bright and, in 1984, an astronomer named James R. Graham detected faint traces of iron in the debris of that supernova. This was the first direct indication that fusion in such a Type I supernova went all the way to iron.

Now it seemed to Graham that a Type I supernova might not be visible at all. If it fused all the way to iron, it would expand hundreds of thousands of times its original diameter so rapidly that its substance would cool down in the process to the point of giving off very little light. Yet the fusion took place, the iron was detected, and there was, despite that, an intense luminosity.

It was Graham's notion that there was some other, slower, source of energy and light, aside from just the fusion. He suggested that the material in the white dwarf fused not into iron-56 (with a nucleus containing 26 protons and 30 neutrons) but into cobalt-56 (with a nucleus containing 27 protons and 29 neutrons).

Whereas the average mass of the 56 particles in iron-56 is, as I said earlier in the article, 0.99884, that of the 56 particles in cobalt-56 is 0.99977. The slight quantity of additional energy in the cobalt-56 is so small that the slope from cobalt-56 to iron-56 is sufficiently gentle to allow the fusion to stall at cobalt-56.

The laws of thermodynamics can't be defeated altogether, however. The cobalt-56 forms, but it can't remain. It is a radioactive nucleus and each one eventually gives off a positron and a gamma ray. The loss of a positron converts a proton to a neutron so that each cobalt-56 nucleus becomes another nucleus with one fewer proton and one additional neutron—in short, a nucleus of iron-56. It is this radioactive change of an entire star's supply of cobalt-56 that provides the energy that produces the luminosity we see in a Type I supernova.

Is there any evidence at all to back up this suggestion? Yes, though the general fusion of nuclei from oxygen all the way up to cobalt may take a matter of only seconds, the decay of cobalt-56 to iron-56 is much more gradual, for cobalt-56 has a half-life of 77 days. If it is the radioactive decay of cobalt-56 that supplies the luminosity of a Type I supernova, then that luminosity ought to decline very regularly just as the radioactivity does. And, apparently, a Type I supernova dims regularly with a half-life of close to 77 days, casting strong suspicion on cobalt-56.

It follows then that although both types of supernovas inject massive nuclei into interstellar matter, the most massive nuclei, such as iron and beyond, are mostly preserved in the shrunken neutron stars and black holes produced by Type II supernovas, but are spread broadcast, along with everything else, by the total explosions of Type I supernovas.

It follows then that most of the iron that has found its way into the Earth's core and its surface rocks—and into our own blood, as well—once existed in white dwarfs that exploded.

15

Opposite!

[NOTE: *This chapter may not seem to belong to this section, but it is the necessary prelude to the following chapter, which does.]*

I spent the last few days in Philadelphia attending the sessions of the annual meeting of the American Association for the Advancement of Science, largely because I was taking part in a symposium on interstellar travel, and because it gives me pleasure, now and then, to wear my scientist hat.

In the course of those days I was interviewed four times and, on one of the occasions, the interviewer said, "But what is antimatter?"

Fortunately, she asked the question of a fellow interviewee, so I let him do the work of explaining and I occupied myself in thinking, with some amusement, of where *I* had first heard of antimatter. It was through my science fiction reading, of course.

In the April 1937 issue of *Astounding Science Fiction,* John D. Clark had a story entitled "Minus Planet," in which an object made of antimatter had blundered into the Solar System and was threatening our planet. That was my first encounter with the concept.

Then, in the August 1937 issue of the same magazine, there was a nonfiction piece by R. D. Swisher, entitled ''What Are Positrons?'' and again I learned about antimatter.

Consequently, in 1939, when I began to write robot stories, I gave my robots ''positronic brains'' as a glamorous science fictional variation of the flat and uninspiring ''electronic brains.''

But when did knowledge of antimatter really start? For that we have to go back to 1928.

In 1928, the English physicist Paul Adrien Maurice Dirac* (1902–84) was studying the electron, one of the only two subatomic particles known at that time, the other being the proton.

For the purpose, Dirac made use of relativistic wave mechanics, the mathematics of which had been worked out by the Austrian physicist Erwin Schrodinger (1887–1961) only two years earlier. In the process, Dirac found that the energy content of a moving electron could be either positive or negative. The positive figure obviously represented the ordinary electron, but, in that case, what did the negative figure (equal in everything but sign) represent?

The easiest way out was to suppose that the negative sign was a mathematical artifact with no meaning in

*Dirac was the son of an immigrant schoolteacher from the French-speaking portion of Switzerland, hence, his name.

physics, but Dirac preferred to find a meaning, if he could.

Suppose that the Universe was made up of a sea of energy levels, with all the negative levels filled with electrons. Above this sea, there is a large but finite number of electrons distributed among the positive energy levels.

If, for some reason, an electron in the sea gains enough energy, it comes popping out of the sea to occupy one of the positive energy levels, and it is then the kind of ordinary electron scientists had grown familiar with. In the sea, however, the departure of the electron leaves a "hole" and this hole behaves as a particle with properties that are the opposite of those of the electron.

Thus, since the electron has an electric charge, that charge had to have been withdrawn from the sea and the hole that appeared must carry a charge of an opposite nature. Since the electron, by a convention that goes back to Benjamin Franklin, is said to have a negative electric charge, the hole must behave as though it has a positive electric charge.

If, then, energy is converted into an electron, the production of the electron must always entail the simultaneous production of a hole, or "antielectron." (The hole is the opposite of an electron and the prefix "anti-" is from a Greek word meaning "opposite.") Dirac was thus predicting "pair-production," the simultaneous production of an electron and antielectron, and it seemed quite clear that you could not produce one of them without the other.

In our section of the Universe, however, a large number of electrons already exists without any sign of an equivalent number of antielectrons. If we accept this fact without questioning the matter too closely, then we can see that once another electron is produced along with its accompanying hole, then, surely, one or another of the many electrons in existence is going to fall into that hole, and do it in a very short time.

Dirac thus predicted that an antielectron was a very short-lived object, which would account for the fact that no one seemed to have encountered one at that time. What's more, Dirac saw that one could not get rid of an antielectron without, at the same time, getting rid of an electron, and vice versa. In other words, you have "mutual annihilation."

In mutual annihilation, the particles must emit, once more, the energy that had been consumed in the pair-production. Mutual annihilation, therefore, had to be accompanied by the production of energetic radiation, or of other particles traveling at high speeds and possessing high kinetic energy, or both.

Since there were only two particles known at the time Dirac worked this out, the negatively charged electron and the positively charged proton, he first wondered if the proton were, by any chance, the antielectron.

Clearly, though, that could not be. In the first place, the proton is 1,836 times as massive as the electron, and it didn't seem at all likely that kicking an electron out of the negative-energy-level sea would produce a hole 1,836 times as massive as that of the extracted particle. The properties of the hole, it seemed logical to suppose, might be opposite in character to those of the extracted particle, but must be the same in quantity.

Thus, the electric charge of the electron is negative so the electric charge of the antielectron should be positive, but the negative charge of one and the positive charge of the other should be precisely the same in quantity. There, at least, the proton fits the bill. Its positive charge is precisely as large as the electron's negative charge.

But this should go for the mass, too, The antielectron might have the same kind of mass the electron does, or perhaps an opposite "antimass," but either way, the mass or antimass should be precisely equal to that of the electron. The proton had the same kind of mass the electron had, but it was widely different in amount.

Furthermore, by Dirac's reasoning, an antielectron ought to be very short-lived and ought to undergo mutual annihilation, almost at once, with any electron it encountered. A proton, however, appeared to be completely stable and showed no tendency whatever to undergo mutual annihilation with electrons.

Dirac, therefore, came to the conclusion that the antielectron was *not* the proton, but was a particle with the mass of an electron and a positive charge.

Still, no one had ever encountered such a positively charged electron, so that most physicists found Dirac's suggestions interesting but insubstantial. They might be merely the speculations of a theorist who was attaching too much literal meaning to mathematical relationships. Until some appropriate observation was made, therefore, Dirac's notions had to be filed under "Interesting, but—"

While Dirac was developing his theory, a Homeric struggle was raging among physicists over the nature of cosmic rays. Some, of whom the American physicist

Robert Andrews Millikan (1868—1953) was the most important, insisted that they were a train of electromagnetic waves, even more energetic and, therefore, shorter in wavelength than gamma rays. Others, of whom the American physicist Arthur Holly Compton (1892–1962) was the most important, insisted that they were a stream of massive, speedy, electrically charged particles. (I won't keep you in suspense. Compton won a complete and unqualified victory.)

In the course of the battle, Millikan had one of his students, Carl David Anderson (1905–   ), study the details of cosmic-ray interaction with the atmosphere. The highly energetic cosmic rays struck the nuclei of atoms in the atmosphere and produced a spray of subatomic particles not much less energetic than the original cosmic rays themselves. It seemed possible that one might argue backward from the particles produced to the nature of the entity that did the producing and thus decide whether the latter was radiational or particulate in nature.

For the purpose, Anderson used a cloud chamber surrounded by a very strong magnetic field. The particles passing through the cloud chamber, which contained gases supersaturated with water vapor, produced charged atom fragments (or "ions") that acted as nuclei for the formation of tiny water droplets. The passage of the particles was thus marked by a thin line of droplets.

What's more, since the particles so detected were electrically charged, their paths (and therefore the lines of droplets) would curve in the presence of a magnetic field. The path of a particle carrying a positive electric charge would curve in one direction; the path of one carrying a negative electric charge would curve in the other direction. The faster the particle, and the more massive, the less it would curve.

The trouble was that the particles produced by cosmic rays smashing into nuclei were so massive or so speedy (or both) that they hardly curved at all. Anderson found there was very little, if anything, he could deduce from their paths.

He therefore had the ingenious idea of putting a lead plate, about 6 millimeters (1/4 inch) thick, across the center of the cloud chamber. Particles smashing into it had more than enough energy to plow right through it. In doing so, however, they would use up quite a bit of their energy and would emerge moving more slowly. They would then curve more sharply and something might be deduced.

In August 1932, Anderson was studying the photographs of various cloud-chamber results he had obtained, and was struck by one in particular. It showed a curved track that looked exactly like the curved tracks produced by speeding electrons.

The track was more curved on one side of the lead plate than on the other. He knew, therefore, that it had entered the chamber on the side of the lesser curvature. It had passed through the lead plate, which slowed it down so that it was more curved on that side. But if it were an electron that had traveled in that direction, it should have curved in the other direction. From its curve, then, Anderson could see at once that he had detected a positively charged electron—the antielectron, in fact.

Naturally, other examples were quickly found, and it was plain that, just as Dirac had predicted, the antielectron did not last long. Within a billionth of a second or so, it would encounter an electron, and mutual annihilation would take place, producing two gamma rays, which were emitted in opposite directions.

Dirac promptly received a Nobel prize for physics in 1933, and Anderson got one in 1936.

One thing about the discovery makes me unhappy. The new particle should have been called the antielectron, as I have been calling it up to now, for that name describes it exactly as "the opposite electron." However, Anderson thought of it as a positive electron. He therefore took the first five letters and the last three letters of the phrase and collapsed it to "positron." That has remained its name ever since.

Of course, if the antielectron is called a positron, then the electron itself ought to be called a "negatron." Then, too, it is not the "-ron" that is the characteristic suffix of subatomic particles, but "-on" as in proton, meson, gluon, lepton, muon, pion, photon, graviton, and so on. If we insist on giving the antielectron a name of its own, then it should be "positon." Indeed, in 1947, there was a move to make use of that name and to call the electron a "negaton," but that failed resoundingly.

It has been "electron" and "positron" ever since, and the two are now unchangeable. But then, science is filled with wrongheaded names wished on it by scientists acting under impulse. (Thus, Murray Gell-Mann invented the ugly term "quark" for the fundamental particles that make up protons. He got it out of *Finnegans Wake,* but that doesn't make it any less ugly. Perhaps he didn't know that in German "quark" means "trash" or "rubbish.")

* * *

Once you have an antielectron, it is impossible to stop there. Dirac's mathematical analysis works precisely the same for protons, for instance, as it does for electrons. Therefore, if there is an antielectron, there should also be an "antiproton."

Yet during the two decades that followed the discovery of the antielectron, there was no sign of an antiproton. Why was this?

No mystery. Mass is a very condensed form of energy, so that it takes a great deal of energy to produce even a small amount of mass. If you want to produce ten times as much mass, you must invest ten times as much energy. The amount of energy required quickly becomes prohibitive.

Since the proton is 1,836 times as massive as an electron, it takes 1,836 times as much energy (all crowded into the kind of tiny volume occupied by a subatomic particle) to produce an antiproton as it does to produce an antielectron.

To be sure, cosmic rays consist of streams of fast-moving massive particles that come in a wide range of energies. Some of the speediest of these particles, and therefore the most energetic, have enough energy and to spare to form proton-antiproton pairs. For that reason, years were spent in studying cosmic-ray events carefully by means of a variety of particle detectors, just in case an antiproton should show up. (Why not? Detect one and you were sure of a Nobel prize.)

One trouble was that as one went up the energy range, the number of cosmic-ray particles possessing that energy decreased. The percentage of cosmic-ray particles possessing enough energy to form a proton-antiproton pair was but a small fraction of the whole. This meant that among the large and complex melange of particles produced by cosmic-ray bombardment, any antiprotons formed would be totally masked by the crowds of others.

Occasionally, someone thought he had detected an antiproton and reported it, but the evidence was never unmistakable. The antiprotons might well have been there, but no one could be sure.

What was needed was a man-made source of energy, one that could be controlled and refined so as to increase the chances of producing and spotting antiprotons. That meant a particle accelerator, one that was more powerful than any built in the 1930s and 1940s.

Finally, in 1954, a particle accelerator was constructed that would produce the necessary energies. This was the Bevatron, built in Berkeley, California. In 1955, the Italian-American physicist Emilio Segrè (1905–) and his American colleague Owen Chamberlain (1920–) worked out a scheme for accomplishing the task.

The plan was to bombard a copper target with very high-energy protons. This should produce proton-antiproton pairs, and a great many other subatomic particles, too. All the particles produced could be made to pass through a strong magnetic field. Protons and other positively charged particles would curve in one direction. Antiprotons and other negatively charged particles would curve in the other direction.

It was calculated that the antiprotons would travel at a certain speed and with a certain curvature. All other negatively charged particles would travel more slowly, or more quickly, and with a different curvature. If, then, a detecting device were placed in some appropriate place and made to operate only at a particular (very short) time after the proton-copper collision, it would be antiprotons and only antiprotons that would be detected. In this way, streams of antiprotons were detected.

Naturally, if antiprotons were produced they would not endure long before encountering the numerous protons that occurred in the Universe all about us. Segrè and Chamberlain allowed the stream of supposed antiprotons which they had detected to strike a piece of glass. Innumerable mutual annihilations took place between the antiprotons in the stream and the protons in the glass.

These annihilations produced particles that could travel through glass more quickly than light. (It is only in a vacuum that the speed of light can't be surpassed.) The particles, as they outpaced light, left a wake of light, called Cherenkov radiation, behind them. The radiation so released was precisely what would be produced by proton-antiproton annihilation.

In both ways, then, by the direct detection of antiprotons and by the study of the radiation produced by annihilation, there was clear evidence that antiprotons had been detected. As a result Segrè and Chamberlain shared the Nobel prize for physics in 1959.

By then, many subatomic particles had been discovered in addition to the electron and proton. Once the antiproton was discovered, it was easy to suppose that opposites would exist for the new particles, too.

That turned out to be right. Every electrically charged particle known has a corresponding particle with a charge opposite to itself. There are "antimuons," "antipions," "antihyperons," "antiquarks," and so on. Every last one of these opposite particles is named by placing the prefix "anti-" before the name of the particle. Only the antielectron is an exception—and the lone exception. It is still called the positron, which must surely annoy anyone who, like myself, values order and method in nomenclature.

All the "anti-" objects can be lumped as "antiparticles."

But what about particles that are *not* charged?

In 1932, the English physicist James Chadwick (1891–1974), discovered the "neutron," which is just a hair more massive than the proton, and differs from that particle in being electrically neutral. (Chadwick received the Nobel prize for physics in 1935 as a result.)

The neutron was found to be the third major component of atoms and of ordinary matter in general. The most common isotope of hydrogen, hydrogen-1, has a single proton as its nucleus, but all other atoms have nuclei made up of both protons and neutrons and these nuclei are accompanied by one or more electrons in the outskirts of the atoms.

No further major components of atoms have ever been found, or are expected to be found. Ordinary matter is made up of protons, neutrons, and electrons and that's all. All other subatomic particles (and there are many) are unstable, high-energy manifestations, or else, if long-lived, exist by themselves and not as part of matter.

What about the neutron, now? An electron is negatively charged while an antielectron is positively charged. A proton is positively charged while an antiproton is negatively charged. The neutron, however, is neutral. It has no charge. What is the opposite of no charge?

Nevertheless, physicists couldn't help thinking there might be an antineutron just the same, even if an electric charge wasn't involved.

Thus, it was reasoned that if a proton and an antiproton whisked by each other in a near-miss, they might not succeed in undergoing mutual annihilation, but they might manage to neutralize each other's electric charge. That would leave two neutral particles that might be opposed to each other in some way; in other words a neutron and an antineutron.

Again, if a neutron and an antineutron are formed, the antineutron ought, soon enough, to collide with a neutron and undergo mutual annihilation, producing particles in some characteristic way.

In 1956, the antineutron was, indeed, discovered, and in 1958 its annihilation reaction was pinpointed. Antiparticles were by then so taken for granted, however, that the discovery of the anti-neutron did not generate a Nobel prize.

And how does the antineutron differ from the neutron? Well, for one thing, although the neutron does not have an overall electric charge, it has something characterized as "spin" that generates a magnetic field. The antineutron has a spin in the opposite direction and a magnetic field that is, therefore, oriented in the direction opposite to that of the neutron.

* * *

In 1965, physicists succeeded in bringing together an antiproton and an antineutron and having them cling together. In ordinary matter, a proton and a neutron, clinging together, make up the nucleus of an atom of hydrogen-2, or "deuterium." What was formed, then, was an "antideuterium" nucleus.

It is clear that an antideuterium nucleus, carrying a negative charge, could easily hold on to a positively charged antielectron. In this way an "antiatom" would be formed. Larger antiatoms could be formed, in principle. The difficulty would consist in forcing all the antiprotons and antineutrons together and, while this was being done, keeping them from undergoing mutual annihilation through random collision with ordinary matter.

We can also imagine antiatoms clinging together to form antimolecules and still larger aggregations. Such aggregations would be "antimatter," though this term could be applied even to antiparticles. —And there you have the answer to the question I described at the beginning of the essay as having been raised by the interviewer.

For a long time, it was supposed that, since particles could not be formed unless accompanied by corresponding antiparticles, there should be as much antimatter in the Universe as matter.

Our Solar System is composed entirely of matter, as otherwise mutual annihilations would occur frequently enough to produce detectable results. Indeed, similar reasoning leads us to be certain that our entire Galaxy is composed of matter only.

Yet might there not be galaxies somewhere composed of antimatter exclusively—"antigalaxies"? It is tempting to suppose that such do exist and are as numerous as galaxies, but the latest theories suggest that particles and antiparticles were not produced in absolutely equal quantities at the time of the big bang. A tiny excess of particles was produced, and this "tiny" excess was still enough to suffice for the building of our vast Universe.

Another question—do all particles, without exception, have antiparticles?

No. A few uncharged particles (not all) are their own antiparticles, so to speak. An example is the photon, which is the unit of all electromagnetic radiation from gamma rays to radio waves, including visible light. The photon is both particle and antiparticle and there is no separate "antiphoton," even in theory.

If there were antiphotons, then the antistars in antigalaxies would emit antiphotons. We could then identify distant objects as antigalaxies by studying the light we receive from them. In actual fact, however, antigalaxies, if there were any, would produce the same light that galaxies do, and photons would be no guide to the existence and location of antigalaxies.

The graviton (which mediates the gravitational interaction) is also its own antiparticle. This means that we can't distinguish between distant galaxies and antigalaxies by any difference in gravitational behavior.

The neutral pion is still another example of a particle that is its own antiparticle.

—And a final question. Might antimatter have some practical use? If not now, then someday?

Let me discuss that aspect of the subject in the next chapter.

16

Sail On! Sail On!

In 1985, with Comet Halley approaching, I was asked by several magazines to do articles on it.

I did such an article for one of them and I received it back with the comment that I had put in all sorts of scientific material of little interest but had neglected the thing most people want to know—when and where to view it best.

I replied by pointing out that it would be useless to do so because the comet was going to pass at quite a distance from the Earth and at such an angle as to be high in the sky only in the Southern Hemisphere. To see it at all, a trip southward would be indicated, which few of the magazine's readers could afford, and if any did go south they would see, at best, a small, dim patch of haze.

I also expressed a little of my bitterness at the incredible ballyhoo and exaggeration that was being displayed in connection with the comet. This was bound to result in the disappointment of a vast number of people and, I said, "I don't intend to add to the hype."

The editor of the magazine was not moved by my eloquence, however. He rejected the article and I did not get a kill fee, either. (However, Gentle Reader, do not weep for me. I sold the article, without changing a word, to another and better magazine for exactly twice the sum the first had offered.)

In January 1985, I had published a book with Walker and Company entitled *Asimov's Guide to Halley's Comet.* In it, I also gave no detailed advice on viewing it. In fact, I stated flatly that the comet would not present a good show. Don't think that some reviewers didn't fault me for omitting detailed information on viewing.

What saddens me about all this is not only that so many people were disappointed in the comet, but that many of them may have been disillusioned with science. I wonder how many of them thought that the dimness of the comet was due to the inefficiency or ignorance of the astronomers arranging the show.

I can only wish that astronomers had been more vocal in their explanation of what the comet's appearance would be like, and a bit readier to denounce all the hucksterism. They were, however, concentrating on the various rocket flybys that were going to (and *did)* make the passage the most successful one ever (scientifically speaking).

But I'm glad it's all over. I did my share of speaking and writing about comets (without hype), even in this essay series, but I'm glad to get on to other subjects. There is, for instance, the matter of interstellar travel, something that is very commonplace in science fiction, but not often discussed elsewhere.

A leading investigator into the possibilities, however, is Dr. Robert L. Forward of Hughes Research Laboratories, who is himself a crackerjack speaker. I had to follow him with a talk of my own at a symposium at a recent meeting of the American Association for the Advancement of Science, and I had to stretch myself to the limit so as not to appear inadequate by contrast.

Let me then take up the subject of interstellar travel, being guided in this by some of Bob's ideas which, of course, I'll expound in my own way.

So far, every vessel we've sent into space, with or without people on board, whether on a suborbital flight, or on a probe to Uranus, has been powered by a chemical reaction engine.

In other words, we have sent out rockets carrying fuel and oxidizer (say, liquid hydrogen and liquid oxygen). When these undergo a chemical reaction, it produces energy which forces the heated gases of the exhaust in one direction while the rest of the rocket moves in the other, in accordance with the law of action and reaction.

The energy of chemical reactions is obtained at the expense of the mass of the system. Mass is a highly concentrated form of energy, and even a very large amount of energy (on the human scale) is formed at the cost of the loss of an insignificant amount of mass.

Thus, suppose we were to burn 1.6 million kilograms (about 1,800 tons) of liquid hydrogen in 12.8 million kilograms (about 14,400 tons) of liquid oxygen, so as to end with 14.4 million kilograms (about 16,200 tons) of water vapor. As a result of some hasty back-of-the-envelope calculations, it seems to me that if we were to weigh the water vapor *precisely,* we would find that it would be just one gram short of the combined original masses of hydrogen and oxygen. All the energy produced by the chemical combination of those millions of tons of hydrogen and oxygen would be equivalent to the loss of one gram of mass. That means that the combination of hydrogen and oxygen releases less than a ten-billionth of its mass in the form of energy.

When you see a tremendous rocket go zooming into the heavens, creating a thunder that makes the earth tremble beneath your feet, just remember, then, that all that fuss represents an insignificant percentage of the energy which, in theory, is present in that mass of fuel and oxidizer.

There may well be some chemicals that on mixing and reacting can outdo hydrogen and oxygen in this respect, but not by much. All chemical fuels are pitiful as sources of energy, and must be accumulated in enormous mass for the amount of energy they can produce. Chemical energy may do very well for ordinary human tasks on Earth's surface. Enough mass can even be accumulated on rocket ships to supply the energy for lofting them into orbit and for exploring the Solar System. For *interstellar* travel, however, chemical reactions are quite hopeless.

The difference between a flight from here to Pluto and one from here to the *nearest* star is about the same as the difference between half a kilometer and the length of Earth's circumference. You can paddle a canoe for half a kilometer but it isn't likely you would consider paddling around the world.

To be sure, a chemical rocket doesn't have to "paddle'' all the way. It can reach a certain speed and then coast, but there will have to be enough fuel to reach that speed and then to decelerate at the other end and, in between, to keep the life-support systems going for the incredible length of time it would take to coast to even the nearest star. It's too much, absolutely too much. The amount of fuel such a ship would have to carry would be simply prohibitive.

Unless there is a richer source of energy than chemical reactions, then, interstellar travel is hopeless.

Nuclear energy was discovered at the opening of the twentieth century. Where chemical energy involves the rearrangements of the electrons in the outer reaches of the atom, nuclear energy involves the rearrangements of the particles within the nucleus. The latter involves much larger energy changes than the former does.

Suppose, then, that instead of burning hydrogen in oxygen, we extract the energy from uranium in the course of its radioactive breakdown. How much uranium would we need to start with in order to have converted 1 gram of mass to energy by the time all of it had turned to lead?

The answer (where's the back of my envelope?) is that about 4,285 grams of uranium, on complete breakdown, will convert 1 of its grams to energy. This means, still, that only 0.023 percent of the mass of the uranium will be converted into energy, but this is a little over three million times as much energy as you would get out of the same mass of hydrogen/oxygen interaction.

There is a catch, though. The radioactive breakdown of uranium, and the consequent production of energy, takes place extraordinarily slowly. Start with 4,285 grams of uranium, and half its breakdown energy would be released only after 4.46 billion years, while 95 percent of its breakdown energy would be produced only after 18 billion years.

Who could wait?

Can the breakdown be speeded up? During the first third of the century, there was no known practical way of doing so. To produce nuclear rearrangements, one had to bombard the nucleus with subatomic particles. This was an enormously inefficient method and the energy that would have to be invested would be many times greater than the energy that could be squeezed out of target nuclei in the process.

It was for this reason that Ernest Rutherford felt there was no hope whatever of being able to make practical use of nuclear energy on a large scale. He described such thoughts as "moonshine." Nor was he a fool. He is one of those on my list of the ten greatest scientists of all times. It was just that he died in 1937 and had no way of foreseeing fission. Had he lived just 2¼ years longer—

Whereas, in natural radioactivity, uranium atoms break down in small bits and pieces, in uranium fission, the atom breaks into two nearly equal pieces. This releases even more energy than ordinary radioactive breakdown.

About 1,077 grams of fissioning uranium will convert 1 of its grams to energy by the time the process is concluded. This means that 0.093 percent of the mass of uranium is converted into energy by fission. This is just about four times as much energy as you can get from an equal weight of uranium undergoing natural radioactive breakdown.

In addition, whereas natural radioactive breakdown cannot be hurried in any practical way, uranium fission can easily be made to take place with explosive speed. Therefore, if we can somehow use nuclear fission to propel spaceships, we will have an energy source some 12 million times as copious as that of chemical interactions. That would surely increase the likelihood of being able to make interstellar voyages, but would it increase it enough?

Bob Forward points out that by using uranium fission to produce the backward thrust of an exhaust, a spaceship could, in 50 years, reach a distance of 200 billion kilometers from the Sun.

This is about 16 times the average distance of Pluto from the Sun, so it isn't bad; but it isn't good, either, for such a distance represents only 1/200 the distance to the *nearest* star. For a ship to take 10,000 years to reach Alpha Centauri certainly leaves a great deal of room for improvement.

But fission isn't the ultimate. Still more energy can be obtained through nuclear fusion. The fusion of four hydrogen nuclei to one helium nucleus is a particularly energy-rich process.

It takes about 146 grams of fusing hydrogen to convert 1 of its grams to energy by the conclusion of fusion. This means that 0.685 percent of the mass of fusing hydrogen is converted into energy, so that there is 7.36 times as much energy to be obtained out of hydrogen fusion as out of uranium fission.

Of course, we don't have controlled fusion as yet, but we do have uncontrolled fusion in the form of hydrogen bombs. People have therefore speculated on the possibility of traveling through space by exploding one hydrogen bomb after another behind the ship.

The debris from the fusion explosions would push outward in all directions and some of it would strike a "pusher plate" attached to the spaceship. The blow is absorbed by means of powerful shock absorbers that would transfer momentum at a reasonable rate to the ship itself.

In 1968, Freeman Dyson imagined an interstellar vessel with a mass of 400,000 tons, carrying 300,000 fusion bombs each weighing one ton. If these bombs were exploded behind the ship at three-second intervals, the ship could be accelerated at 1 g.—That is, everyone on board would feel an apparent normal gravitational pull in the direction of the exploding bombs. The ship would be rising like a steadily accelerating elevator and this acceleration would push your feet against the "floor"—actually the rear—of the ship.

In ten days, the 300,000 fusion bombs would have been consumed and the ship would have attained a speed of about 10,000 kilometers per second. If the ship is aimed in the right direction and coasts at this speed, it will pass Alpha Centauri in 130 years. If one wants to make a landing on some object orbiting about one of the stars of that system, one would naturally have to carry another 300,000 hydrogen bombs and explode them in front of the ship—or else turn the ship around with ordinary chemical reaction motors, and then explode the hydrogen bombs behind the ship again, that behind now facing Alpha Centauri.

Reaching Alpha Centauri in 130 years is far better than reaching it in 10,000 years, but it still means that the original voyagers would have to spend their entire life on board ship, and that their great-grandchildren, most likely, would be landing somewhere in the Alpha Centauri planetary systems. What's more, we can't count on relativistic effect making the time seem shorter for the crew. Even at 10,000 kilometers per second (one-thirtieth the speed of light) relativistic effects are insignificant. The apparent time for the crew members will be cut by an hour or so, no more.

Things would be better, perhaps, if we had controlled fusion, and could maintain such reactions on board ship for an extended period. The products of the fusion reaction could be allowed to leak out behind at a steady and manageable rate, producing a jet that could accelerate the ship in the other direction, exactly like a rocket exhaust. In this way, then, all the energy of fusion could be directed into acceleration, instead of only that part of the power of the exploding bombs that happened to be directed toward the pusher plate, while the power in other directions is wasted on the vacuum of space.

Then, too, the controlled fusion reaction would supply the power continuously, instead of doing so by way of a series of successive shocks. Nevertheless, I don't think the time taken to reach Alpha Centauri would be reduced below the century mark.

Besides, even hydrogen fusion converts less than 1 percent of the fuel into energy. Is there any way we can do still better?

Yes, there is such a thing as antimatter. (I discussed this in the previous chapter.)

Antimatter will combine with matter and will, in the process, annihilate all the matter interacting. Thus half a gram of antimatter, combining with half a gram of matter, would produce 146 times as much energy as fusing a gram of hydrogen, or 1,075 times as much energy as fissioning a gram of uranium, or several billion times as much energy as burning a gram of hydrogen in oxygen.

The most easily available form of antimatter is the antielectron (or positron). If, however, antielectrons are allowed to interact with electrons, then they produce pure energy in the form of gamma-ray photons. These emerge in all directions and cannot easily be channeled into an exhaust.

Next simplest is the antiproton, which is the nucleus of an antihydrogen atom, whereas the proton is the nucleus of a hydrogen atom. For simplicity's sake, then, we can speak of antihydrogen and hydrogen.

If antihydrogen and hydrogen are allowed to interact, a mixture of unstable particles, "pions" and "antipions," are the major products. These, being electrically charged, can be channeled into a very rapid rocket exhaust, driving the ship forward. The pions and antipions turn into "muons" and "antimuons" after a brief interval, and, after a somewhat longer interval, the muons and antimuons turn into electrons and antielectrons. In the end, all of the mass of the original hydrogen and antihydrogen is converted into energy, except for the trace amount that escapes in the form of electrons and non-electrons that manage to remain apart and not interact.

In addition, a large quantity of ordinary hydrogen can be added to the interacting mixture. This hydrogen would be heated to enormous temperatures and would also emerge as a rocket exhaust, so to speak, adding to the acceleration.

Forward has calculated that 9 kilograms of antihydrogen and 4 tons of hydrogen, between them, could accelerate a spaceship to one-tenth the speed of light (30,000 kilometers per second) and that will mean reaching Alpha Centauri in something like 40 years.

Perhaps, if enough antimatter is used, speeds equal to one-fifth the speed of light (60,000 kilometers per second) can be reached. In that case, a round trip to Alpha Centauri might take no more than 40 years. It would then be possible to go there and back in a single lifetime, and we can imagine that, assuming spaceships are large enough and sufficiently comfortable, some young people might be willing to devote their lives to it.

—But there are difficulties.

To begin with, in our section of the Universe, and perhaps in the Universe as a whole, antiprotons exist only in the tiniest traces. It would be necessary to manufacture them.

This can be done, for instance, by bombarding metal targets with high-speed protons. The spray of energy that results is converted to particles in part, these particles including some antiprotons. At the moment, the number of antiprotons formed is only about 2 for every 100 million protons fired at the target. To try to gather enough antiprotons for an interstellar mission at this rate would be an expensive undertaking indeed, but it is natural to hope that the efficiency of antiproton production would be greatly increased in time.

Once the antiprotons are produced, another problem arises. Antiprotons will immediately react with any protons they encounter and every bit of ordinary matter contains protons. The task of keeping hydrogen and oxygen from uncontrollable explosion before you are ready for an orderly burning is as nothing compared with the task of keeping antiprotons from premature explosion of a far more drastic sort.

Once formed, antiprotons must be isolated from all matter, and kept isolated till one is ready for the interaction with protons. While this is difficult, it is not impossible. We can imagine solid antihydrogen being stored in a vacuum, the "walls" of which consist of electric or magnetic fields. If this is done someday, ships powered by antihydrogen may streak through space from Earth to Mars in weeks, to Pluto in months, and to the nearest star in decades.

In everything I have described so far, the interstellar spaceships must carry fuel. The most concentrated possible fuel we know of is the antiproton, but what if you needed no fuel at all?

You would need none if the fuel existed all about you in space and, in a way, it does. Space is not truly empty; not even between galaxies, and certainly not between the stars within a galaxy. There is a scattering of hydrogen atoms (or their nuclei) everywhere.

Suppose you launch a ship with a minimum of ordinary fuel, just enough to build up the speed that would allow you to scoop up enough interstellar hydrogen. Such hydrogen could then be made to fuse and the products of fusion fired out behind as exhaust, first supplementing and then replacing the original supply of fuel.

You can then continue to accelerate indefinitely, because there is no danger of running out of fuel, and the faster you go, the more fuel you can collect in a unit of time. This is an "interstellar ramjet" and, using it, you can finally achieve speeds as near the speed of light as you wish. Allowing for acceleration and deceleration, you might make the round trip to Alpha Centauri in only 15 years.

That's what it would take as far as people on Earth are concerned. The astronauts, themselves, would, at ultrafast speeds, experience a slowdown in apparent time passage. What might seem like 15 years to the stay-at-home on Earth, might seem only about 7 years to the astronauts.

Seven years out of a lifetime isn't at all bad. It's only twice the time spent, by the survivors of Magellan's voyage, nearly five centuries ago, to circumnavigate the Earth for the first time.

More than that, if you continue coasting at very nearly the speed of light, hardly any additional time will pass as far as the astronauts are concerned. If they decide to travel to the other end of the Galaxy, or to an alien galaxy a hundred million light-years away, they may feel the passage only of some additional months in the first case, and a couple of additional years in the second.

Of course, they would come home to find that a hundred thousand or a hundred million years had passed on Earth, which might rather spoil their fun. Still, with interstellar ramjets, the problem of travel among the stars might seem to be solved.

But there are catches. In order to collect enough hydrogen from interstellar space, assuming it contains 1,000 atoms per cubic centimeter, you would need a scoop that is over a hundred kilometers across, and that supposes that the hydrogen atoms are ionized and therefore carry an electric charge so that they can be scooped up by appropriate electric or magnetic fields.

Unfortunately the interstellar space about the Sun is sparse in hydrogen, and contains less than 0.1 hydrogen atoms for every cubic centimeter. For that reason the scoop would have to be 10,000 kilometers across, and have an area about two-fifths that of Earth's surface. What's more, the hydrogen atoms in our vicinity are not ionized and are, therefore, not easily collectible. (Perhaps that is not entirely unfortunate, however. If our neighborhood of space were thick with ionized hydrogen, we would be near enough to something violent to make it a little dubious as to whether life could survive on Earth.)

Besides, even if we could scoop up enough hydrogen and push it through the fusion engines, it may not be practical for an interstellar spaceship to go faster than a fifth the speed of light.

After all, the faster we go, the more difficult it is to avoid collisions with small objects and the more damage such collision will wreak. Even if we are fortunate enough to miss all sizable objects, we can scarcely expect to miss the dust and individual atoms that are scattered throughout space.

At two-tenths the speed of light, dust and atoms might not do significant damage even in a voyage of 40 years, but the faster you go, the worse it is—space begins to become abrasive. When you begin to approach the speed of light, each hydrogen atom becomes a cosmic-ray particle, and they will fry the crew. (A hydrogen atom or its nucleus striking the ship at nearly the speed of light is a cosmic-ray particle, and there is no difference if the ship strikes a hydrogen atom or a hydrogen nucleus at nearly the speed of light. As Sancho Panza says: "Whether the stone strikes the pitcher, or the pitcher strikes the stone, it is bad for the pitcher." So 60,000 kilometers per second may be the *practical* speed limit for space travel.

Even the interstellar ramjet makes use of the rocket principle. Bob Forward, however, talks of "rocketless rocketry." The ship might be pushed by pellets fired from within the Solar System or by a maser beam or a laser beam.

These devices would also make it unnecessary for an interstellar spaceship to carry its own fuel and would also allow a buildup to speeds near that of light. The advantage of such things over ramjets would be that they would not depend on surrounding space having very special and hard-to-meet characteristics.

Of course, the technical difficulties would be formidable indeed. A laser beam, for instance, would have to strike an aluminum-film sail that would be 1,000 kilometers in diameter and, even if extremely thin, would be sure to weigh 80 million kilograms or so. —And speeds above one-fifth that of light would remain impractical.

I think, then, that a 40-year round trip, with antimatter fuel, is the best we can do if we want to explore the interstellar spaces within the lifetime of an astronaut. And even that will take him only to the nearest star.

That is not to be sneezed at, of course. It will allow us to study, in detail, a second star that is very much like our Sun (Alpha Centauri A), one that is distinctly smaller and dimmer (Alpha Centauri B), and one that is a small red dwarf (Alpha Centauri C)—to say nothing of any planetary objects that may circle any of the three.

If we can establish an independent civilization in the Alpha Centauri system, it would be able to send ships in the direction away from us, reaching a star in an astronaut's lifetime that we could not.

In this way, a wave of exploration can leapfrog outward in all directions, each new base being able to reach one or two or even three stars that others might not be able to get to. Humanity could then spread throughout the Galaxy over the space of several hundreds of thousands of years.

Contact need not involve travel alone. Each new world can maintain contact with nearby worlds by means of signals that travel at the speed of light. News can travel from one world to another in relays, and pass from one end of the galaxy to another in a hundred thousand years or so.

All that, however, is not the sort of interstellar travel, or the sort of Galactic Empire, that we science fiction writers are constantly describing.

No, for what *we* want there simply must be faster-than-light travel. Nothing else will do. That has been a staple of science fiction ever since E. E. Smith introduced it in *The Skylark of Space* when it was published in

1928. Since then everyone, including me, has used it (with or without some plausible explanation).

Unfortunately, since I see nothing on the horizon that has much of a practical chance of giving us faster-than-light speed, I'm afraid that my Galactic Empire of the Foundation series is likely forever to remain—science fiction.

Nevertheless, I warn you, I have every intention of continuing to use it, just the same.

Part IV
Something Extra

17

The Relativity of Wrong

I received a letter from a reader the other day. It was handwritten in crabbed penmanship so that it was very difficult to read. Nevertheless, I tried to make it out just in case it might prove to be important.

In the first sentence, he told me he was majoring in English Literature, but felt he needed to teach me science. (I sighed a bit, for I knew very few English Lit majors who are equipped to teach me science, but I am very aware of the vast state of my ignorance and I am prepared to learn as much as I can from anyone, however low on the social scale, so I read on.)

It seemed that in one of my innumerable essays, here and elsewhere, I had expressed a certain gladness at living in a century in which we finally got the basis of the Universe straight.

I didn't go into detail in the matter, but what I meant was that we now know the basic rules governing the Universe, together with the gravitational interrelationships of its gross components, as shown in the theory of relativity worked out between 1905 and 1916. We also know the basic rules governing the subatomic particles and their interrelationships, since these are very neatly described by the quantum theory worked out between 1900 and 1930. What's more, we have found that the galaxies and clusters of galaxies are the basic units of the physical Universe, as discovered between 1920 and 1930.

These are all twentieth-century discoveries, you see.

The young specialist in English Lit, having quoted me, went on to lecture me severely on the fact that in *every* century people have thought they understood the Universe at last, and in *every* century they were proved to be wrong. It follows that the one thing we can say about our modern "knowledge" is that it is *wrong*.

The young man then quoted with approval what Socrates had said on learning that the Delphic oracle had proclaimed him the wisest man in Greece. "If I am the wisest man," said Socrates, "it is because I alone know that I know nothing." The implication was that I was very foolish because I was under the impression I knew a great deal.

Alas, none of this was new to me. (There is very little that is new to me; I wish my correspondents would realize this.) This particular thesis was addressed to me a quarter of a century ago by John Campbell, who specialized in irritating me. He also told me that all theories are proven wrong in time.

My answer to him was, "John, when people thought the Earth was flat, they were wrong. When people thought the Earth was spherical, they were wrong. But if *you* think that thinking the Earth is spherical is *just as wrong* as thinking the Earth is flat, then your view is wronger than both of them put together."

The basic trouble, you see, is that people think that "right" and "wrong" are absolute; that everything that isn't perfectly and completely right is totally and equally wrong.

However, I don't think that's so. It seems to me that right and wrong are fuzzy concepts, and I will devote this essay to an explanation of why I think so.

First, let me dispose of Socrates because I am sick and tired of this pretense that knowing you know nothing is a mark of wisdom.

No one knows *nothing.* In a matter of days, babies learn to recognize their mothers.

Socrates would agree, of course, and explain that knowledge of trivia is not what he means. He means that in the great abstractions over which human beings debate, one should start without preconceived, unexamined notions, and that he alone knew this. (What an enormously arrogant claim!)

In his discussions of such matters as "What is justice?" or "What is virtue?" he took the attitude that he knew nothing and had to be instructed by others. (This is called "Socratic irony," for Socrates knew very well that he knew a great deal more than the poor souls he was picking on.) By pretending ignorance, Socrates lured others into propounding their views on such abstractions. Socrates then, by a series of ignorant-sounding questions, forced the others into such a mélange of self-contradictions that they would finally break down and admit they didn't know what they were talking about.

It is the mark of the marvelous toleration of the Athenians that they let this continue for decades and that it wasn't till Socrates turned seventy that they broke down and forced him to drink poison.

Now where do we get the notion that "right" and "wrong" are absolutes? It seems to me that this arises in the early grades, when children who know very little are taught by teachers who know very little more.

Young children learn spelling and arithmetic, for instance, and here we tumble into apparent absolutes.

How do you spell sugar? Answer: s-u-g-a-r. That is *right.* Anything else is *wrong.*

How much is 2 + 2? The answer is 4. That is *right.* Anything else is *wrong.*

Having exact answers, and having absolute rights and wrongs, minimizes the necessity of thinking, and that pleases both students and teachers. For that reason, students and teachers alike prefer short-answer tests to essay tests; multiple-choice over blank short-answer tests; and true-false tests over multiple-choice.

But short-answer tests are, to my way of thinking, useless as a measure of the student's understanding of a subject. They are merely a test of the efficiency of his ability to memorize.

You can see what I mean as soon as you admit that right and wrong are relative.

How do you spell "sugar?" Suppose Alice spells it p-q-z-z-f and Genevieve spells it s-h-u-g-e-r. Both are wrong, but is there any doubt that Alice is wronger than Genevieve? For that matter, I think it is possible to argue that Genevieve's spelling is superior to the "right" one.

Or suppose you spell "sugar": s-u-c-r-o-s-e, or $C_{12}H_{22}O_{11}$. Strictly speaking, you are wrong each time, but you're displaying a certain knowledge of the subject beyond conventional spelling.

Suppose then the test question was: how many different ways can you spell "sugar." Justify each.

Naturally, the student would have to do a lot of thinking and, in the end, exhibit how much or how little he knows. The teacher would also have to do a lot of thinking in the attempt to evaluate how much or how little the student knows. Both, I imagine, would be outraged.

Again, how much is 2 + 2? Suppose Joseph says: 2 + 2 = purple, while Maxwell says: 2 + 2 = 17. Both are wrong but isn't it fair to say that Joseph is wronger than Maxwell?

Suppose you said: 2 + 2 = an integer. You'd be right, wouldn't you? Or suppose you said: 2 + 2 = an even integer. You'd be rather righter. Or suppose you said: 2 + 2 = 3.999. Wouldn't you be *nearly* right?

If the teacher wants 4 for an answer and won't distinguish between the various wrongs, doesn't that set an unnecessary limit to understanding?

Suppose the question is, how much is 9 + 5?, and you answer 2. Will you not be excoriated and held up to ridicule, and will you not be told that 9 + 5 = 14?

If you were then told that 9 hours had passed since midnight and it was therefore 9 o'clock, and were asked what time it would be in 5 more hours, and you answered 14 o'clock on the grounds that 9 + 5 = 14, would you not be excoriated again, and told that it would be 2 o'clock? Apparently, in that case, 9 + 5 = 2 after all.

Or again suppose, Richard says: 2 + 2 = 11, and before the teacher can send him home with a note to his mother, he adds, "To the base 3, of course." He'd be right.

Here's another example. The teacher asks: "Who is the fortieth President of the United States?" and Barbara says, "There isn't any, teacher."

"Wrong!" says the teacher, "Ronald Reagan is the fortieth President of the United States."

"Not at all," says Barbara, "I have here a list of all the men who have served as President of the United States under the Constitution, from George Washington to Ronald Reagan, and there are only thirty-nine of them, so there is no fortieth President."

"Ah," says the teacher, "but Grover Cleveland served two nonconsecutive terms, one from 1885 to 1889, and the second from 1893 to 1897. He counts as both the twenty-second and twenty-fourth President. That is why Ronald Reagan is the thirty-ninth person to serve as President of the United States, and is, at the same time, the fortieth President of the United States."

Isn't that ridiculous? Why should a person be counted twice if his terms are nonconsecutive, and only once if he served two consecutive terms. Pure convention! Yet Barbara is marked wrong—just as wrong as if she had said the fortieth President of the United States is Fidel Castro.

Therefore, when my friend the English Literature expert tells me that in every century scientists think they have worked out the Universe and are *always wrong*, what

I want to know is *how* wrong are they? Are they always wrong to the same degree? Let's take an example.

In the early days of civilization, the general feeling was that the Earth was flat.

This was not because people were stupid, or because they were intent on believing silly things. They felt it was flat on the basis of sound evidence. It was *not* just a matter of "That's how it looks," because the Earth does *not* look flat. It looks chaotically bumpy, with hills, valleys, ravines, cliffs, and so on.

Of course, there are plains where, over limited areas, the Earth's surface *does* look fairly flat. One of those plains is in the Tigris-Euphrates area where the first historical civilization (one with writing) developed, that of the Sumerians.

Perhaps it was the appearance of the plain that may have persuaded the clever Sumerians to accept the generalization that the Earth was flat; that if you somehow evened out all the elevations and depressions, you would be left with flatness. Contributing to the notion may have been the fact that stretches of water (ponds and lakes) looked pretty flat on quiet days.

Another way of looking at it is to ask what is the "curvature" of Earth's surface. Over a considerable length, how much does the surface deviate (on the average) from perfect flatness. The flat-Earth theory would make it seem that the surface doesn't deviate from flatness at all, that its curvature is 0 to the mile.

Nowadays, of course, we are taught that the flat-Earth theory is *wrong*; that it is all wrong, terribly wrong, absolutely. But it isn't. The curvature of the

Earth is *nearly* 0 per mile, so that although the flat-Earth theory is wrong, it happens to be *nearly* right. That's why the theory lasted so long.

There were reasons, to be sure, to find the flat-Earth theory unsatisfactory and, about 350 B.C., the Greek philosopher Aristotle summarized them. First, certain stars disappeared beyond the Southern Hemisphere as one traveled north, and beyond the Northern Hemisphere as one traveled south. Second, the Earth's shadow on the Moon during a lunar eclipse was always the arc of a circle. Third, here on Earth itself, ships disappeared beyond the horizon hull-first in whatever direction they were traveling.

All three observations could not be reasonably explained if the Earth's surface were flat, but could be explained by assuming the Earth to be a sphere.

What's more, Aristotle believed that all solid matter tended to move toward a common center, and if solid matter did this, it would end up as a sphere. A given volume of matter is, on the average, closer to a common center if it is a sphere than if it is any other shape whatever.

About a century after Aristotle, the Greek philosopher Eratosthenes noted that the Sun cast a shadow of different lengths at different latitudes (all the shadows would be the same length if the Earth's surface were flat). From the difference in shadow length, he calculated the size of the earthly sphere and it turned out to be 25,000 miles in circumference.

The curvature of such a sphere is about 0.000126 per mile, a quantity very close to 0 per mile as you can see, and one not easily measured by the techniques at the disposal of the ancients. The tiny difference between 0 and 0.000126 accounts for the fact that it took so long to pass from the flat Earth to the spherical Earth.

Mind you, even a tiny difference, such as that between 0 and 0.000126, can be extremely important. That difference mounts up. The Earth cannot be mapped over large areas with any accuracy at all if the difference isn't taken into account and if the Earth isn't considered a sphere rather than a flat surface. Long ocean voyages can't be undertaken with any reasonable way of locating one's own position in the ocean unless the Earth is considered spherical rather than flat.

Furthermore, the flat Earth presupposes the possibility of an infinite Earth, or of the existence of an "end" to the surface. The spherical Earth, however, postulates an Earth that is both endless and yet finite, and it is the latter postulate that is consistent with all later findings.

So although the flat-Earth theory is only slightly wrong and is a credit to its inventors, all things considered, it is wrong enough to be discarded in favor of the spherical-Earth theory.

And yet is the Earth a sphere?

No, it is *not* a sphere; not in the strict mathematical sense. A sphere has certain mathematical properties—for instance, all diameters (that is, all straight lines that pass from one point on its surface, through the center, to another point on its surface) have the same length.

That, however, is not true of the Earth. Various diameters of the Earth differ in length.

What gave people the notion the Earth wasn't a true sphere? To begin with, the Sun and the Moon have outlines that are perfect circles within the limits of measurement in the early days of the telescope. This is consistent with the supposition that the Sun and Moon are perfectly spherical in shape.

However, when Jupiter and Saturn were observed by the first telescopic observers, it became quickly apparent that the outlines of those planets were not circles, but distinct ellipses. That meant that Jupiter and Saturn were not true spheres.

Isaac Newton, toward the end of the seventeenth century, showed that a massive body would form a sphere under the pull of gravitational forces (exactly as Aristotle had argued), but only if it were not rotating. If it were rotating, a centrifugal effect would be set up which would lift the body's substance against gravity, and this effect would be greater the closer to the equator you progressed. The effect would also be greater the more rapidly a spherical object rotated and Jupiter and Saturn rotated very rapidly indeed.

The Earth rotated much more slowly than Jupiter or Saturn so the effect should be smaller, but it should still be there. Actual measurements of the curvature of the Earth were carried out in the eighteenth century and Newton was proved correct.

The Earth has an equatorial bulge, in other words. It is flattened at the poles. It is an "oblate spheroid" rather than a sphere. This means that the various diameters of the Earth differ in length. The longest diameters are any of those that stretch from one point on the equator to an opposite point on the equator. This "equatorial diameter" is 12,755 kilometers (7,927 miles). The shortest diameter is from the North Pole to the South Pole and this "polar diameter" is 12,711 kilometers (7,900 miles).

The difference between the longest and shortest diameters is 44 kilometers (27 miles), and that means that the "oblateness" of the Earth (its departure from true sphericity) is $^{44}/_{12,755}$, or 0.0034. This amounts to $^{1}/_{3}$ of 1 percent.

To put it another way, on a flat surface, curvature is 0 per mile everywhere. On Earth's spherical surface, curvature is 0.000126 per mile everywhere (or 8 inches per mile). On Earth's oblate spheroidical surface, the curvature varies from 7.973 inches to the mile to 8.027 inches to the mile.

The correction in going from spherical to oblate spheroidal is much smaller than going from flat to spherical. Therefore, although the notion of the Earth as sphere is wrong, strictly speaking, it is not *as* wrong as the notion of the Earth as flat.

Even the oblate-spheroidal notion of the Earth is wrong, strictly speaking. In 1958, when the satellite *Vanguard 1* was put into orbit about the Earth, it was able to measure the local gravitational pull of the Earth—and therefore its shape—with unprecedented precision. It turned out that the equatorial bulge south of the equator was slightly bulgier than the bulge north of the equator, and that the South Pole sea level was slightly nearer the center of the Earth than the North Pole sea level was.

There seemed no other way of describing this than by saying the Earth was pearshaped and at once many people decided that the Earth was nothing like a sphere but was shaped like a Bartlett pear dangling in space. Actually, the pearlike deviation from oblate-spheroid perfect was a matter of yards rather than miles and the adjustment of curvature was in the millionths of an inch per mile.

In short, my English Lit friend, living in a mental world of absolute rights and wrongs, may be imagining that because all theories are *wrong,* the Earth may be thought spherical now, but cubical next century, and a hollow icosahedron the next, and a doughnut shape the one after.

What actually happens is that once scientists get hold of a good concept they gradually refine and extend it with greater and greater subtlety as their instruments of measurement improve. Theories are not so much wrong as incomplete.

This can be pointed out in many other cases than just the shape of the Earth. Even when a new theory seems to represent a revolution, it usually arises out of small refinements. If something more than a small refinement were needed, then the old theory would never have endured.

Copernicus switched from an Earth-centered planetary system to a Sun-centered one. In doing so, he switched from something that was obvious to something that was apparently ridiculous. However, it was a matter of finding better ways of calculating the motion of the planets in the sky and, eventually, the geocentric theory was just left behind. It was precisely because the old theory gave results that were fairly good by the measurement standards of the time that kept it in being so long.

Again, it is because the geological formations of the

Earth change *so* slowly and the living things upon it evolve *so* slowly that it seemed reasonable at first to suppose that there was *no* change and that Earth and life always existed as they do today. If that were so, it would make no difference whether Earth and life were billions of years old or thousands. Thousands were easier to grasp.

But when careful observation showed that Earth and life were changing at a rate that was very tiny but *not* zero, then it became clear that Earth and life had to be very old. Modern geology came into being, and so did the notion of biological evolution.

If the rate of change were more rapid, geology and evolution would have reached their modern state in ancient times. It is only because the difference between the rate of change in a static Universe and the rate of change in an evolutionary one is that between zero and very nearly zero that the creationists can continue propagating their folly.

Again, how about the two great theories of the twentieth century; relativity and quantum mechanics?

Newton's theories of motion and gravitation were very close to right, and they would have been absolutely right if only the speed of light were infinite. However, the speed of light is finite, and that had to be taken into account in Einstein's relativistic equations, which were an extension and refinement of Newton's equations.

You might say that the difference between infinite and finite is itself infinite, so why didn't Newton's equations fall to the ground at once? Let's put it another way, and ask how long it takes light to travel over a distance of a meter.

If light traveled at infinite speed, it would take light 0 seconds to travel a meter. At the speed at which light actually travels, however, it takes it 0.0000000033 seconds. It is that difference between 0 and 0.0000000033 that Einstein corrected for.

Conceptually, the correction was as important as the correction of Earth's curvature from 0 to 8 inches per mile was. Speeding subatomic particles wouldn't behave the way they do without the correction, nor would particle accelerators work the way they do, nor nuclear bombs explode, nor the stars shine. Nevertheless, it was a tiny correction and it is no wonder that Newton, in his time, could not allow for it, since he was limited in his observations to speeds and distances over which the correction was insignificant.

Again, where the prequantum view of physics fell short was that it didn't allow for the "graininess" of the Universe. All forms of energy had been thought to be continuous and to be capable of division into indefinitely smaller and smaller quantities.

This turned out to be not so. Energy comes in quanta, the size of which is dependent upon something called Planck's constant. If Planck's constant were equal to 0 erg-seconds, then energy would be continuous, and there would be no grain to the Universe. Planck's constant, however, is equal to 0.0000000000000000000000000066 erg-seconds. That is indeed a tiny deviation from zero, so tiny that ordinary questions of energy in everyday life need not concern themselves with it. When, however, you deal with subatomic particles, the graininess is sufficiently large, in comparison, to make it impossible to deal with them without taking quantum considerations into account.

* * *

Since the refinements in theory grow smaller and smaller, even quite ancient theories must have been sufficiently right to allow advances to be made; advances that were not wiped out by subsequent refinements.

The Greeks introduced the notion of latitude and longitude, for instance, and made reasonable maps of the Mediterranean basin even without taking sphericity into account, and we still use latitude and longitude today.

The Sumerians were probably the first to establish the principle that planetary movements in the sky exhibit regularity and can be predicted, and they proceeded to work out ways of doing so even though they assumed the Earth to be the center of the Universe. Their measurements have been enormously refined but the principle remains.

Newton's theory of gravitation, while incomplete over vast distances and enormous speeds, is perfectly suitable for the Solar System. Halley's Comet appears punctually as Newton's theory of gravitation and laws of motion predict. All of rocketry is based on Newton, and *Voyager II* reached Uranus within a second of the predicted time. None of these things were outlawed by relativity.

In the nineteenth century, before quantum theory was dreamed of, the laws of thermodynamics were established, including the conservation of energy as first law, and the inevitable increase of entropy as the second law. Certain other conservation laws such as those of momentum, angular momentum, and electric charge were also established. So were Maxwell's laws of electromagnetism. All remained firmly entrenched even after quantum theory came in.

Naturally, the theories we now have might be considered wrong in the simplistic sense of my English Lit correspondent, but in a much truer and subtler sense, they need only be considered incomplete.

For instance, quantum theory has produced something called "quantum weirdness" which brings into serious question the very nature of reality and which produces philosophical conundrums that physicists simply can't seem to agree upon. It may be that we have reached a point where the human brain can no longer grasp matters, or it may be that quantum theory is incomplete and that once it is properly extended, all the "weirdness" will disappear.

Again, quantum theory and relativity seem to be independent of each other, so that while quantum theory makes it seem possible that three of the four known interactions can be combined into one mathematical system, gravitation—the realm of relativity—as yet seems intransigent.

If quantum theory and relativity can be combined, a true "unified field theory" may become possible.

If all this is done, however, it would be a still finer refinement that would affect the edges of the known—the nature of the big bang and the creation of the Universe, the properties at the center of black holes, some subtle points about the evolution of galaxies and supernovas, and so on.

Virtually all that we know today, however, would remain untouched and when I say I am glad that I live in a century when the Universe is essentially understood, I think I am justified.

ABOUT THE AUTHOR

Born in Petrovichi, Russia, in 1920, Isaac Asimov emigrated to New York with his parents in 1923. It was during his childhood in Brooklyn that he developed his lifelong love of books by reading the volumes in his local library, shelf by shelf. Accepted to Columbia University at the age of fifteen, Asimov earned his Bachelor of Science degree, and went on to receive his Ph.D. in chemistry in 1948. He taught biochemistry at the Boston University School of Medicine until 1958, when he turned to writing full-time.

Dr. Asimov began writing science fiction at the age of eleven, and had his first short story published in 1938. His first book-length work of science fiction, *Pebble in the Sky,* was published by Doubleday in 1950, and he soon branched out into nonfiction as well. Asimov has written on nearly every subject under the sun, ranging from math and physics to the Bible and Shakespeare, and currently has over 365 published books to his credit. He has received numerous honorary degrees and writing awards, including a special Hugo Award honoring his internationally bestselling *Foundation Trilogy* as "The Best All-Time Science Fiction Series." And, he was recently named Grandmaster of Science Fiction by the Science Fiction Writers of America.

Not content with specializing in his first field of expertise—the sciences—Dr. Asimov has proved a successful essayist, mystery writer, editor, journalist, biographer, humorist, and all-around exponent of the written word. Happiest working in the seclusion of his two-room office, lined with more than 2,000 books, Asimov lives in Manhattan with his wife, writer Janet O. Jeppson.